The Universe Without a Creator

How Cosmology Explains the Cosmos Without God

(Book 3)

Dr Prabhat Das

MD, PhD, FACP

No God Required

How Science Explains Reality Without the Divine

Hardcover ISBN: 979-8-951067-06-7

Paperback ISBN: 979-8-951067-05-0

eBook ISBN: 979-8-951067-04-3

Printed in the United States of America

TABLE OF CONTENTS:

Author's Note

Why I Wrote This Series

Some books begin in libraries. Some begin in classrooms. This series began much earlier, in the mind of a curious boy growing up in India.

I was born into a Hindu religious family. Religion was not a distant theory in my childhood. It was part of daily life. It was in prayers, rituals, festivals, stories, customs, family conversations, and the quiet assumptions that surrounded me. God was not discussed as a possibility. God was treated as reality.

My father was a firm believer in an omnipotent God. To him, God was real, powerful, protective, and deeply involved in human life. His belief was sincere. It was part of his strength, his discipline, and his view of the world. I grew up respecting that seriousness.

But from my teenage years, another force began to grow inside me. Curiosity.

I wanted to know why things happened. I wanted to understand nature, disease, death, the stars, the human body, and the mind. I was not satisfied with an answer simply because it was old, sacred, or repeated by many people. I wanted to know whether it was true.

In high school, I became deeply impressed by the lives and writings of great scientists. I still remember going to the local library and borrowing books written by, or about, Newton, Darwin, Einstein, and others. Those books were not popular among students at that time. In fact, I may have been the first student to borrow some of them. They were covered with dust. Some had even been damaged by insects.

But to me, they were treasures. Those dusty books opened a new world.

Newton showed me that nature follows laws.

Darwin showed me that life has a history far deeper than ancient stories.

Einstein showed me that reality is stranger, deeper, and more beautiful than ordinary imagination.

These thinkers did not ask me to believe blindly. They invited me to understand. They did not frighten me with punishment. They opened the door to wonder. They showed me that truth does not become weaker when questioned. It becomes stronger.

That invitation changed my life.

My curiosity did not end with school. It continued through medical college, scientific training, clinical practice, and decades of work as a physician. In medicine, I learned the value of evidence. A diagnosis cannot be accepted because it feels comforting. A treatment cannot be trusted because it is traditional. A belief about the body must face reality. It must face observation, testing, correction, and outcome.

That same habit of mind gradually shaped the way I looked at religion & God.

Even now, at retirement age, I continue to read almost every day. Science, philosophy, history, religion, psychology, evolution, cosmology, neuroscience, and the human mind still fascinate me. The more I read, the more I see how far human knowledge has come. Many things once explained by God are now understood through natural causes. Thunder, disease, eclipse, famine, fertility, mental illness, death, life, and the vast universe itself no longer need supernatural explanation.

This book series, "**No God Required: How Science Explains Reality Without the Divine", is a 7 book series**, and is my attempt to share that long journey in clear and simple language.

I do not write these books to insult believers. Many believers are kind, sincere, moral, generous, and deeply thoughtful people. I know this from my own family, my own culture, my patients, my friends, and my life experience. Faith often gives people comfort, hope, discipline, community, and emotional strength.

But faith does not decide truth.

A belief may comfort us and still need examination. A tradition may be ancient and still be mistaken. A prayer may bring peace and still not prove that anyone is listening. A sacred story may be beautiful and still be human-made.

My goal is to understand how belief began, why it became powerful, and why it still feels so real to so many people.

Why did human beings create gods?

Why did religion become one of the strongest forces in history?

Why do beliefs learned in childhood feel like truth?

Why do different cultures create different gods?

Why did science gradually make divine explanations unnecessary?

Can human beings live meaningful, moral, compassionate, and courageous lives without God?

These are serious questions. They deserve serious answers.

This series is based on knowledge gathered from many great books, scholars, scientists, historians, philosophers, psychologists, and thinkers. I have tried to bring their insights together in language that an ordinary educated reader can understand. My hope is that these books will encourage readers to think more rationally, question inherited beliefs, respect evidence, and study these subjects more deeply.

I also hope they will show that life without God does not have to be empty, cold, selfish, or immoral.

A life guided by reason, compassion, truth, science, justice, and human responsibility can be deeply meaningful. We do not need divine command to care for one another. We do not need fear of hell to be moral. We do not need supernatural stories to feel wonder before the universe.

Human beings are already capable of love, courage, kindness, beauty, and truth.

The universe is already astonishing.

Life is already precious.

Our responsibility to one another is already urgent.

No God is required for that.

Introduction: The Last Great Hiding Place of God

The Question Beneath the Stars

Long before there were temples, scriptures, priests, telescopes, satellites, or equations, human beings stood under the night sky and looked upward.

They saw darkness filled with light.

They saw the Moon change shape.

They saw stars return in silent patterns.

They saw comets appear without warning.

They saw eclipses turn day into fear.

They heard thunder. They saw lightning. They depended on rain. They feared drought, disease, death, famine, and mystery.

They did not know galaxies.

They did not know gravity.

They did not know nuclear fusion.

They did not know space-time.

They did not know cosmic microwave background radiation.

They did not know that Earth is one small planet orbiting one ordinary star in one galaxy among billions.

But they felt awe.

They felt fear.

They felt small.

And when human beings feel small before something vast and mysterious, they often imagine a mind behind it.

That is how the sky became sacred.

For thousands of years, the heavens seemed to declare divine power. The universe looked like a creation. The stars looked like signs. The Sun looked like a god or a gift from God. The Moon seemed to carry meaning. Earth felt central. Human life felt watched.

Even today, that old feeling has not disappeared.

A person may accept modern medicine, use a smartphone, fly in airplanes, trust electricity, and know that Earth orbits the Sun. Yet when that same person stands beneath the stars and asks, "Where did all this come from?" the ancient answer often returns:

God.

God becomes the maker of the universe.

God becomes the cause of the Big Bang.

God becomes the mind behind mathematics.

God becomes the lawgiver behind natural law.

God becomes the designer behind the constants of physics.

God becomes the reason there is something rather than nothing.

This is why the universe remains one of the last great hiding places of God.

Science has already explained much that religion once claimed. Thunder no longer needs divine anger. Lightning no longer needs a heavenly weapon. Eclipses no longer need cosmic warning. Disease no longer needs demons or punishment. Fossils no longer need flood myths. Planetary motion no longer needs angelic guidance. Life no longer needs special creation.

But the universe itself still feels different.

It feels too large.

Too deep.

Too mysterious.

Too close to the final question.

This book begins there.

It asks one clear question:

Does the universe really need a creator?

When God Moved to the Edge of Knowledge

The first book in this series examined why human beings created God. It showed that gods grew from fear, death, nature, dreams, grief, tribe, authority, ritual, and the human hunger for meaning. Before science, God gave people an answer when they had no better answer. God gave comfort when life was dangerous. God gave order when the world felt chaotic.

The second book showed what happened when science began to explain what God once explained.

The sky became astronomy.

The body became anatomy and physiology.

Matter became chemistry.

Earth became geology.

Life became evolution.

Disease became microbiology.

The mind became psychology and neuroscience.

In one field after another, God became unnecessary as an explanation.

This did not mean belief disappeared. It did not. God remained in prayer, grief, music, ritual, identity, family, fear, hope, and culture. But as an explanation of the natural world, God steadily lost ground.

Then God moved.

That movement is one of the most important patterns in the history of religion.

When God was no longer needed to explain lightning, God was placed behind electricity.

When God was no longer needed to explain disease, God was placed behind biology.

When God was no longer needed to explain evolution, God was placed behind evolution.

When God was no longer needed to explain planetary motion, God was placed behind gravity.

When science explained more and more inside the universe, God was moved to the universe itself.

God became the creator.

God became the first cause.

God became the source of natural law.

God became the explanation for why anything exists at all.

This is where many modern arguments for God now stand.

Not in the storm.

Not in the plague.

Not in the orbit of Mars.

Not in the fossil record.

Not in the human eye.

But at the beginning of the universe.

This book asks whether that final move succeeds.

Why the Universe Feels Like the Strongest Argument

The universe feels different from other questions.

A fever can be explained by infection.

An eclipse can be explained by the positions of the Sun, Moon, and Earth.

A fossil can be explained by ancient life, sediment, minerals, pressure, and time.

A planet's motion can be explained by gravity and space-time.

But the universe itself seems to raise a deeper question.

Why is there a universe at all?

Why is there space?

Why is there time?

Why is there matter?

Why are there laws?

Why does mathematics describe nature so well?

Why are the constants of physics compatible with stars, atoms, chemistry, planets, and life?

Why is there something rather than nothing?

These questions are not trivial. They deserve respect.

A serious atheist should not wave them away with arrogance. A serious scientific thinker should not pretend that every cosmic mystery has been solved. It has not. Modern cosmology is powerful, but not complete. Physics has explained much, but not everything.

Dark matter remains mysterious.

Dark energy remains mysterious.

Quantum gravity remains unfinished.

The earliest state of the universe remains difficult.

The deepest nature of time, law, and existence remains under investigation.

This book will not pretend otherwise.

But honesty about mystery does not help religion as much as many believers think.

A mystery is not a creator.

An unanswered question is not evidence.

A frontier of physics is not a temple.

The correct answer to what we do not know is not "therefore God."

The correct answer is, "We do not yet know."

That sentence is not weakness.

It is intellectual honesty.

Science-based atheism does not require pretending that every mystery has vanished. It requires a simpler discipline: do not turn mystery into God without evidence.

That discipline is one of the central lessons of this book.

The Big Bang Is Not a Creation Story

Many people hear the words Big Bang and imagine a cosmic beginning that confirms religion.

They picture empty darkness. Then God speaks. A flash appears. Space explodes. Matter bursts outward. Stars and galaxies eventually form. Human beings arrive. The ancient creation story seems to return, dressed in scientific language.

This is a misunderstanding.

The Big Bang is not God lighting a cosmic fuse.

It is not an explosion inside empty space.

It is not the first sentence of Genesis translated into physics.

It is not proof of the Bible, the Quran, the Vedas, or any scripture.

The Big Bang is a scientific model. It describes the observable universe expanding and cooling from an earlier hot, dense state. It is supported by evidence, including cosmic expansion, galactic redshift, cosmic microwave background radiation, light element abundance, and large-scale structure.

That makes it powerful.

But it does not make it religious.

The Big Bang tells us that the universe has a physical history. It tells us that the cosmos was once hotter, denser, and very different from what we see today. It tells us that galaxies, stars, planets, atoms, and eventually life belong to a long natural story.

But it does not say, "God did it."

It does not identify a creator.

It does not prove a personal God.

It does not prove a loving God.

It does not prove a prayer-answering God.

It does not prove heaven, hell, miracles, divine judgment, scripture, or life after death.

Even if the universe had a beginning in some sense, a beginning is not automatically a God.

A beginning is a question.

God is a proposed answer.

And proposed answers need evidence.

The First Cause Does Not Automatically Become God

One of the oldest arguments for God is the first-cause argument.

Everything that begins must have a cause.

The universe began.

Therefore, the universe must have a cause.

That cause is God.

At first, this sounds simple and powerful. It has emotional force because human beings are used to causes. A child exists because parents existed. A house exists because builders built it. A book exists because someone wrote it. A fire begins because something ignites it.

So we ask:

What caused the universe?

But the question is not as simple as it sounds.

Cause is a concept we learn from events inside the universe. One thing happens, then another thing follows. A match is struck, then a flame appears. A stone is thrown, then glass breaks. A virus enters the body, then illness may develop.

But when we ask about the cause of the universe as a whole, we are no longer asking an ordinary question.

The universe is not one object sitting inside a larger room.

The universe is the physical system that includes space, time, matter, energy, and natural law as we know them.

If time itself is part of the universe, then what does it mean to ask what happened before time? If cause normally means one event before another, what happens to cause when before itself becomes uncertain?

This does not prove there was no cause.

But it does show that the simple religious argument is not simple at all.

Even if a first cause were accepted, another problem remains.

Why call it God?

A first cause, by itself, does not give us a personal God. It does not give us a God who loves, forgives, judges, listens to prayer, writes scriptures, performs miracles, chooses nations, or offers heaven.

At most, it gives a philosophical starting point.

Religion then adds enormous claims to that starting point.

This book will examine that move carefully.

A first cause is not automatically God.

And God is not an explanation simply because we give mystery a sacred name.

The Problem of Who Created God

If everything needs a cause, why does God not need one?

This question is often asked by children, but it is not childish. It is one of the most serious questions in the entire debate about God.

The usual religious answer is that God is eternal, uncaused, necessary, or self-existent. God does not need a cause because God is not like ordinary things.

But this answer raises another question.

If God can be uncaused, why cannot the deepest reality of the universe be uncaused?

If God can be necessary, why cannot the cosmos, physical law, quantum reality, or some deeper natural foundation be necessary?

If God can be eternal, why cannot some form of natural reality be eternal, even if our observable universe has a finite history?

The believer may say, "God is different."

Perhaps. But that difference must be shown, not merely asserted.

To say "everything needs a cause except God" is not an argument. It is an exemption. It protects God from the very rule used to prove God.

This is a major weakness in many creator arguments.

God is introduced to explain the mystery of the universe. Yet God is often more mysterious than the universe. God is said to be timeless, spaceless, immaterial, all-powerful, all-knowing, conscious, intentional, morally perfect, and capable of creating reality itself.

How does this make the mystery smaller?

If we are allowed to stop explanation somewhere, why not stop with reality itself, or with a deeper natural foundation still being studied?

Why stop with God?

The question is not whether the universe is mysterious.

It is.

The question is whether God explains the mystery better.

Very often, God does not explain the mystery.

God only renames it.

Fine-Tuning: A Serious Question, Not a Proof

Some religious arguments are weak because they depend on ignorance that science has already corrected.

Fine-tuning is different.

It deserves more careful attention.

The universe appears to have certain features that make stars, atoms, chemistry, planets, and life possible. If some constants or early conditions were very different, life as we know it might not exist. This has led many people to argue that the universe looks delicately arranged for life.

The religious conclusion is simple:

The universe is fine-tuned.

Fine-tuning requires a tuner.

Therefore, God designed the universe.

This argument can feel powerful.

It appears more scientific than older arguments. It does not depend on denying evolution or germ theory. It uses physics, cosmology, and mathematics. It accepts that the universe has laws, then asks why those laws seem life-permitting.

This book will treat the argument seriously.

But serious does not mean successful.

Fine-tuning raises questions, but it does not prove God.

We do not know whether the constants of nature could have been different in the way the argument assumes. We do not know the full probability distribution of possible universes. We do not know whether deeper laws constrain the values we observe. We do not know whether a multiverse exists. We do not know whether life could exist in forms very different from life as we know it.

We also know that most of the universe is not friendly to life.

Space is deadly. Stars burn. Radiation kills. Most planets are lifeless. Vast regions are empty, frozen, airless, crushed, irradiated, or hostile. If the universe was designed specifically for human beings, it is strange that nearly all of it is instantly fatal to human beings.

Fine-tuning is a serious mystery.

But a serious mystery is still not proof.

It is a question to investigate, not a conclusion to worship.

The Anthropic Correction

The anthropic principle begins with a simple observation:

We can only observe a universe compatible with observers.

This may sound obvious, but it matters.

If the universe could not produce observers, no one would be here to notice it. We should not be surprised to find ourselves in conditions where beings like us can exist, because if those conditions did not exist, we would not be here.

This does not solve everything. It does not explain why there is a universe. It does not explain why there are laws. It does not fully explain fine-tuning.

But it weakens a common religious mistake.

Religion often says:

The universe fits us, therefore it was made for us.

Science replies:

We fit this universe because we arose within it.

That is a very different interpretation.

We are not outside reality, inspecting it as a product made for us. We are inside reality. We are made of its matter, shaped by its laws, and limited by its conditions. Our existence is not independent of the universe. It is one of the universe's outcomes.

The universe may not be fitted to us.

We may be fitted to this universe.

That idea is less flattering than religion.

But it is more honest.

The Multiverse and the Discipline of Humility

Some scientists and philosophers have suggested that our observable universe may not be all that exists. There may be many cosmic domains, perhaps with different conditions, different histories, or different values of physical constants.

This idea is called the multiverse.

It is fascinating. It is controversial. It is not proven. It should not be presented as settled fact. It should not become an atheist myth replacing a religious myth.

But it matters.

If many cosmic domains exist, then it becomes less surprising that at least one has conditions compatible with observers. We would naturally find ourselves in such a region because we could not exist in a region where observers cannot arise.

This does not prove the multiverse.

But it does show that divine design is not the only possible response to fine-tuning.

There may be deeper laws. There may be selection effects. There may be multiple cosmic domains. There may be physical explanations not yet discovered. There may be things about the constants that we do not yet understand.

The honest conclusion is not "therefore God."

The honest conclusion is "keep investigating."

That is the difference between science and theology.

Theology often rushes to final answers.

Science can live with unfinished questions.

This humility is not weakness.

It is strength.

Quantum Cosmology and the Failure of Common Sense

Human common sense is useful.

It helps us cross streets, cook food, avoid danger, raise children, treat patients, and live in the ordinary world. But common sense was not designed to understand the earliest universe.

In ordinary life, time moves forward. Causes come before effects. Objects exist in space. Nothing means empty absence. A thing cannot come from nothing. These ideas feel obvious because they work in daily experience.

But modern physics has repeatedly shown that daily experience is not a reliable guide to the deepest reality.

Earth feels still, but it moves.

The Sun appears to circle Earth, but it does not.

Time feels absolute, but relativity changed that.

Solid matter feels completely solid, but its atomic structure is far stranger.

At very small scales, quantum physics does not behave like ordinary objects.

So when people say, "The universe must have a cause because everything needs a cause," they may be taking common sense far beyond its proper territory.

Near the earliest universe, our ordinary ideas of time, cause, beginning, and nothing may not apply in simple ways.

This does not mean science has solved everything.

It has not.

But it does mean that religious arguments based on everyday intuition are not as strong as they first appear.

The universe is not obligated to obey human common sense.

The mistake of religion has often been to take human imagination and project it onto reality.

Science teaches a harder discipline:

Let reality correct imagination.

Stephen Hawking and the End of Cosmic Ignition

Stephen Hawking became one of the most famous scientists in the world because he brought cosmic questions into public life.

He asked questions that seemed almost religious in scale.

Did the universe have a beginning?

What is time?

What happens inside black holes?

Can the universe be self-contained?

Does physics leave any role for a creator?

In *A Brief History of Time*, Hawking explored the nature of time, black holes, singularities, and the origin of the universe. Later, in *The Grand Design*, he and Leonard Mlodinow argued that physical law may be enough to explain why a universe can exist.

That claim became famous and controversial.

It should be handled carefully.

Hawking did not settle every question. He did not prove, in a final philosophical sense, that no God can exist. He did not fully explain why there are laws of physics. He did not make all metaphysical questions disappear.

But he did something extremely important.

He challenged the idea that God is needed to start the universe.

He helped millions of readers see that cosmic origin may be approached through law, mathematics, and scientific theory rather than divine command.

The believer may still ask, "Where did the laws come from?"

That is a fair question.

But the atheist may answer, "Where did God come from?"

And more importantly:

If physical law can explain more of cosmic origin than divine will, then God again becomes unnecessary as an explanation.

This is the central thread of the whole series.

Science explains.

God retreats.

Why "God Did It" Is Not Enough

At the heart of this book is one simple question:

What does "God did it" actually explain?

If we ask why the universe exists and someone says, "God created it," what have we learned?

Do we know the mechanism?

Do we know how a timeless mind creates time?

Do we know how an immaterial being creates matter?

Do we know how divine intention becomes physical law?

Do we know why God chose this universe rather than another?

Do we know why the universe contains so much emptiness, radiation, violence, death, and waste?

Do we know how to test the claim?

Do we know what observation would prove it wrong?

Do we know whether the creator is Yahweh, Allah, Brahman, a deistic intelligence, a cosmic mind, or something else?

Usually, no.

"God did it" sounds like an answer because it is grammatically complete.

But scientifically, it is empty.

A real explanation should do work. It should clarify. It should connect. It should predict. It should help us understand why this happened rather than something else. It should be open to evidence. It should be capable of correction.

God does none of this in cosmology.

God can be used to explain any universe.

A small universe? God made it.

A vast universe? God made it.

A universe with life? God made it.

A universe mostly hostile to life? God made it.

A universe with laws? God made the laws.

A universe with mystery? God's ways are mysterious.

An explanation that fits every possible outcome explains no specific outcome.

That is why “God did it” is not enough.

It is not a real explanation.

It is a stopping point.

Science refuses to stop there.

What This Book Does Not Claim

This book is strong in its conclusion, but it must remain honest in its method.

It does not claim that science has fully explained why there is something rather than nothing.

It does not claim that cosmology has solved every mystery of the earliest universe.

It does not claim that dark matter and dark energy are fully understood.

It does not claim that the multiverse is proven.

It does not claim that Stephen Hawking ended all debate about God.

It does not claim that every religious person is ignorant.

It does not claim that the universe is emotionally easy to accept without God.

The universe is still mysterious.

Human beings are still small.

Death is still real.

Meaning still matters.

Wonder still matters.

But none of this proves God.

This book makes a more careful and more powerful claim:

God is not needed as an explanation for what cosmology has successfully explained, and God has not been shown to explain what cosmology has not yet explained.

That is enough.

Science-based atheism does not require false certainty. It requires disciplined honesty. It says: where evidence explains, accept evidence. Where mystery remains, do not invent supernatural certainty.

That is a stronger position than faith.

Faith often claims to know what it does not know.

Science admits what it does not know and keeps searching.

Atheism Beneath the Stars

Some people think atheism makes the universe empty.

They imagine that without God, the stars are only burning gas, life is only chemistry, love is only biology, and human beings are accidental dust. They think a universe without God must be cold, meaningless, and morally dead.

This is wrong.

A universe without God is not empty.

It is real.

The stars are not less beautiful because they are made of plasma and powered by nuclear fusion. Galaxies are not less magnificent because gravity shapes them. The cosmic microwave background is not less astonishing because it is radiation. The laws of physics are not less elegant because no divine lawgiver is required.

Understanding does not destroy wonder.

It deepens wonder.

A child may think thunder is the voice of God. Science shows it is atmospheric electricity. The truth does not make the storm less powerful. It makes our understanding more honest.

A believer may think stars are lamps placed by God. Science shows they are suns, born from gas and gravity, burning across cosmic time. That does not make the night sky less beautiful. It makes it more vast, more ancient, and more moving.

Science-based atheism is not the death of wonder.

It is wonder without illusion.

It is awe without worship.

It is humility without surrendering the mind.

It is honesty before reality.

The universe does not need to be created for us in order to be magnificent.

The Road Ahead

This book will move step by step.

First, we will look at the universe as modern science reveals it: vast, ancient, dynamic, and not human-centered.

Then we will see how the sacred sky became the physical cosmos.

We will meet Einstein and see how space, time, gravity, and matter became part of one natural framework.

We will meet Hubble and see how galaxies and redshift revealed an expanding universe.

We will examine the Big Bang, not as a religious creation event, but as a scientific model of cosmic expansion from an earlier hot, dense state.

We will look at the evidence for Big Bang cosmology, including the cosmic microwave background and the large-scale structure of the universe.

Then we will turn to philosophy.

Does the universe need a first cause?

Does the first cause prove God?

Who created God?

Can something be necessary without being divine?

Then we will examine fine-tuning.

Do the constants of nature point to design?

Does the anthropic principle weaken that argument?

Could multiverse ideas change the question?

What should honest scientific humility look like?

Then we will move deeper into quantum cosmology and the limits of common sense.

Finally, we will examine Stephen Hawking and the claim that the universe may not require God as creator.

At the end, we will return to the main question:

Does the universe need God?

The answer developed in this book will be clear.

The universe remains mysterious.

But God is not required.

The Question This Book Asks

The question of this book is not whether the universe makes us feel awe.

It does.

The question is not whether the universe is mysterious.

It is.

The question is not whether some people feel God when they look at the stars.

They do.

The question is more precise:

Does the universe require God as an explanation?

That is the question.

And the answer, when examined through science, logic, and philosophical care, is no.

The Big Bang does not prove God.

A first cause does not prove the God of religion.

Fine-tuning does not prove design.

Natural law does not require a divine lawgiver.

Mystery does not justify supernatural belief.

The unknown is not owned by religion.

The universe is not evidence that God exists.

It is evidence that reality is larger, stranger, deeper, and more magnificent than ancient human imagination knew.

God was once placed in the sky.

Then the sky became astronomy.

God was placed behind the planets.

Then the planets became physics.

God was placed behind cosmic beginnings.

Now cosmology asks whether even that hiding place is necessary.

This book argues that it is not.

The universe does not need a creator in order to exist as a subject of awe, study, and truth.

It does not need divine management to be lawful.

It does not need supernatural purpose to be meaningful to us.

It does not need God to be real.

And reality, honestly understood, is enough.

Sources

Stephen Hawking, *A Brief History of Time*

Stephen Hawking and Leonard Mlodinow, *The Grand Design*

Sean Carroll, *The Big Picture*

Steven Weinberg, *The First Three Minutes*

Carl Sagan, *Cosmos*

Brian Greene, *The Fabric of the Cosmos*

Martin Rees, *Just Six Numbers*

David Hume, *Dialogues Concerning Natural Religion*

Stanford Encyclopedia of Philosophy, "Cosmological Argument"

Stanford Encyclopedia of Philosophy, "Fine-Tuning"

Chapter 1: The Universe Is Not a Human-Centered Stage

The Small Blue Planet

Look at Earth from space.

No borders are visible.

No temples.

No flags.

No scriptures.

No armies.

No arguments about which god owns which land, which people, or which truth.

There is only a small blue-white sphere floating in darkness.

Clouds swirl over oceans. Continents appear beneath a thin, fragile layer of air. Everything human beings have ever loved, feared, built, worshiped, destroyed, defended, or dreamed has happened on that small world.

Now pull farther away.

Earth becomes smaller. The Moon appears beside it, gray and lifeless. The Sun becomes the center of the solar system. Planets move in their paths. Some are rocky. Some are gas giants. Some have rings. Some have many moons. Between them stretches enormous emptiness.

Pull farther again.

The Sun becomes one star among hundreds of billions in the Milky Way. The solar system almost disappears. The star that gives us daylight, warmth, crops, seasons, rain cycles, and life is no longer a divine lamp hung above Earth. It is one star in one galaxy.

Pull farther still.

The Milky Way becomes one galaxy among billions. Galaxies gather into clusters. Clusters form long filaments across space. Between them stretch vast dark voids. The universe expands beyond ordinary imagination.

Now ask the old religious question:

Was all this built mainly for us?

Was this immense cosmos created so that one species of primate, on one small planet, around one ordinary star, could pray, sin, worship, doubt, and be judged?

This question matters because modern astronomy changed more than our map of the sky. It changed the emotional foundation of human-centered religion.

The universe is not a small sacred theater.

It is not a house built around human life.

It is not a cosmic courtroom with Earth at the center and heaven just above the clouds.

It is vast, ancient, dynamic, violent, mostly empty, and mostly hostile to human life.

This does not make human life meaningless.

It makes our understanding more honest.

The first lesson of modern cosmology is humility. Human beings are not at the center of the universe. Earth is not the main stage of cosmic history. The stars were not placed in the sky merely as signs for us. The galaxies do not exist to decorate our nights.

The universe is larger than ancient religion imagined.

Once we understand that, the old idea of a human-centered creation begins to weaken.

The Ancient Human-Centered World

For most of human history, people naturally imagined the universe around themselves.

This was understandable. Human beings lived on Earth. They looked upward and saw the Sun, Moon, planets, and stars move across the sky. The ground beneath their feet felt still. The heavens appeared to revolve around them. Rain came from above. Thunder came from above. Light came from above. The dead were buried below. The gods were often imagined above.

The universe seemed layered.

Heaven above.

Earth here.

The underworld below.

Human beings in the middle.

This structure felt emotionally satisfying. It gave people a place in reality. It made the universe feel morally arranged. It made human life seem central to cosmic purpose.

In many religious traditions, the world was not merely a planet. It was the stage of divine drama. God created. God commanded. Human beings obeyed or disobeyed. Sin entered. Judgment followed. Salvation was offered. Prayer rose upward. Blessings came down. The universe became a moral theater.

This picture was comforting.

It told people that their lives mattered to the ruler of the cosmos. It told them that their suffering was seen. It told them that their prayers were heard. It told them that obedience and disobedience had eternal importance.

But comfort is not evidence.

The fact that an idea gives meaning does not make it true.

The old human-centered cosmos was born from limited knowledge. Ancient people did not know Earth was a planet. They did not know the Sun was a star. They did not know the Milky Way was a galaxy. They did not know there were billions of galaxies. They did not know that the universe is about 13.8 billion years old. They did not know that most of cosmic history passed before human beings appeared.

They placed themselves at the center because they did not yet know the size of the stage.

Science changed that.

Earth Lost Its Throne

Earth feels important because it is our home.

Everything we love happens here. Birth happens here. Childhood happens here. Language, family, memory, music, medicine, science, grief, justice, injustice, love, and death happen here. Every human life ever lived has unfolded on this small planet.

Emotionally, Earth is everything to us.

Cosmically, Earth is tiny.

It is one planet orbiting one star. It is not the largest planet even in its own solar system. It is not at the center of the solar system. It is not at the center of the galaxy. It is not at the center of the universe.

This loss of centrality began with Copernicus and continued through modern astronomy. At each stage, human beings were moved farther from the center.

First, Earth was no longer the center.

Then the Sun was no longer unique.

Then the Milky Way was no longer the whole universe.

Then galaxies became countless.

Then the universe became not only vast, but expanding.

This history matters because religion often depends emotionally on human importance. It tells us that the universe was created with human beings in mind. It says that the creator of all reality watches us, judges us, loves us, forgives us, and listens to our prayers.

But the physical universe does not look arranged around us.

It looks indifferent to us.

Earth is precious to us because we live here, not because the universe points toward Earth as its purpose.

A believer may say, "God can still care about one small planet."

That is possible as belief.

But possible is not the same as proven.

The smallness of Earth does not disprove every possible idea of God. But it weakens the old religious feeling that the universe is obviously centered on human beings.

The evidence does not point to cosmic privilege.

It points to cosmic humility.

The Sun Became a Star

The Sun was once surrounded by worship, fear, and sacred imagination.

This is easy to understand. Without the Sun, there is no human life. It gives light. It gives warmth. It drives climate. It supports photosynthesis. It shapes seasons. It marks time. It rises each morning with a power that ancient people could easily feel as divine.

They did not know nuclear fusion.

They did not know that the Sun is one star among many.

They did not know that stars are born, burn fuel, change, and die.

They saw the Sun as a heavenly power.

Science made the Sun natural.

The Sun is a star. It is magnificent, but not unique. It is not the greatest star, the oldest star, the largest star, or the center of all things. It is one ordinary star in one region of one galaxy.

This does not make the Sun less wonderful.

It makes its wonder more truthful.

The Sun is not a god.

It is a physical object.

It shines because of nuclear fusion in its core. Hydrogen is converted into helium. Energy is released. Light and heat travel outward. Some of that energy reaches Earth and makes life possible.

This explanation is more powerful than worship because it tells us what is actually happening.

A solar god may inspire hymns.

Physics explains the star.

This is the pattern of science-based atheism throughout this series. The world becomes more understandable when we stop treating natural forces as divine. We do not lose wonder. We lose illusion.

The Sun is not proof of God.

It is proof that nature is capable of extraordinary power without requiring supernatural intention.

The Milky Way Lost Its Uniqueness

For a long time, human beings did not know what the Milky Way was.

It appeared as a pale band across the night sky, a river of light, a sacred path, a mystery. Different cultures gave it different meanings. Some saw it as a road of souls. Some saw it as spilled milk. Some saw it as a heavenly river. Ancient imagination turned the Milky Way into story because the human mind naturally turns mystery into meaning.

Modern astronomy revealed something far more astonishing.

The Milky Way is a galaxy.

It contains hundreds of billions of stars, along with gas, dust, dark matter, planets, stellar remnants, and immense structure. Our Sun is only one star within it.

For a while, even this seemed enormous enough.

Then came a greater shock.

The Milky Way is not the universe.

It is one galaxy among billions.

This discovery changed the scale of reality. The human mind struggles to absorb it. A billion is already difficult to imagine. Billions of galaxies, each with billions of stars, are almost beyond ordinary thought.

What does this mean for religion?

It does not logically disprove every possible God. A believer may say God created all galaxies. That is possible as belief. But the old emotional structure was weakened again.

If the purpose of the cosmos was human salvation, why such vastness?

If the stars were made as signs for human beings, why billions of galaxies beyond human sight?

If Earth is the moral center of creation, why did nearly all cosmic reality play no role in human history?

If human beings are the main purpose of creation, why did the universe exist for billions of years before we appeared?

These questions do not prove atheism by themselves. But they make the old religious picture less convincing.

The universe looks less like a house built for humans and more like a vast natural reality in which humans appeared very late.

The Observable Universe and the Larger Unknown

It is important to be precise.

When scientists speak of the observable universe, they mean the part of the universe from which light has had time to reach us. Because the universe has a finite age and light travels at a finite speed, we cannot see everything that may exist.

There is a cosmic horizon.

Beyond that horizon, there may be more universe. Perhaps far more. We do not know its full extent. Our observable universe may be only one region of a much larger reality.

This matters because ancient religion was created within a tiny field of vision. Human beings built gods from the world they could see: the sky above, the land below, the river, the storm, the Sun, the Moon, the stars, birth, disease, death, and dream.

Modern science has shown that the visible world is not the whole of reality.

Even with powerful telescopes, we see only what the laws of light and cosmic history allow us to see. The universe is larger than direct human experience, larger than ancient scripture, larger than religious imagination, and perhaps larger than the observable region itself.

This should make us humble.

It should also make us careful.

When a religious tradition claims that the creator of all reality revealed final truth to one ancient tribe, one prophet, one book, or one culture on one small planet, we should ask whether that claim fits the scale of what we now know.

The universe is not only vast.

It is larger than any human-centered story.

Cosmic Time and Human Time

Religion often compresses time into human meaning.

Creation, ancestors, prophets, lawgivers, sin, judgment, salvation, and final destiny are placed in a timeline that feels close to human life. Sacred stories often make the world feel young, intimate, and morally focused.

Cosmic time is different.

The universe is about 13.8 billion years old.

Earth is about 4.54 billion years old.

Life on Earth is ancient, but human beings are extremely recent.

Written history is only a tiny moment.

Modern science has stretched time far beyond sacred imagination. The universe existed for billions of years before Earth formed. Earth existed for billions of years before human beings appeared. Stars lived and died before our solar system was born. Heavy elements were forged in earlier generations of stars. The atoms in our bodies come from cosmic processes older than our planet.

This is a magnificent thought.

It is also humbling.

Human beings did not appear at the beginning. We arrived late. We are not the opening chapter of the universe. We are a recent development in a long cosmic story.

This matters because many religions place human beings near the center of divine purpose. The universe is created, humanity appears, moral drama begins, and cosmic destiny revolves around human obedience, sin, faith, and salvation.

But cosmic time tells a different story.

For most of its history, the universe had no human beings. No prayer. No scripture. No temples. No theology. No atheism. No religion at all.

The universe was already ancient before the first human mind wondered about God.

This does not make human life worthless. It makes human life precious in a different way. We are not precious because the universe was built around us. We are precious because, for a brief moment, matter became aware of itself.

That is a deeper and more honest wonder.

A Universe Mostly Hostile to Life

If the universe was designed mainly for human beings, it is strange that nearly all of it kills human beings instantly.

Space has no breathable air. Extreme cold dominates many regions. Stars burn with unimaginable heat. Black holes crush matter. Radiation damages life. Most planets are lifeless. Many are frozen, scorched, toxic, airless, or violently unstable. Even Earth, our home, is safe only in a thin layer near its surface.

Human beings survive in a very narrow range of conditions.

We need oxygen, water, pressure, temperature control, food, protection from radiation, and a stable environment. Remove any of these, and life quickly becomes impossible.

The universe is not a garden built for human comfort.

It is a vast and mostly deadly reality in which a tiny region happens to support life.

This matters when people say the universe is fine-tuned for us.

Perhaps certain physical conditions allow life to exist. That is true and important. But we should not confuse "life is possible somewhere" with "the universe was made for us."

Most of the universe is not friendly to life.

Even on Earth, life survives within danger. Earthquakes, volcanoes, storms, floods, droughts, disease, predators, parasites, genetic errors, cancer, and extinction are all part of nature. Life is not protected by cosmic kindness. It struggles within physical conditions.

A loving designer is not the simplest explanation for a universe that is overwhelmingly hostile to the beings supposedly at its center.

A natural universe explains this better.

Life appears where conditions allow it. It does not appear where conditions do not allow it. Human beings survive because Earth has suitable conditions, not because the whole universe is arranged for our comfort.

This is not despair.

It is clarity.

The Universe Does Not Flatter Us

Religion often flatters human beings.

It tells us that the creator of the universe made us in His image. It tells us that our prayers are heard by the ruler of reality. It tells us that our moral choices are watched by heaven. It tells us that our species has cosmic importance.

Modern cosmology does not flatter us.

It tells us that Earth is small.

It tells us that the Sun is ordinary.

It tells us that the Milky Way is one galaxy among billions.

It tells us that human beings appeared very late.

It tells us that the universe is not arranged visibly around our needs.

This can feel emotionally difficult.

But truth is often less flattering than illusion.

Science-based atheism asks us to accept reality without demanding that reality make us central. It asks us to grow beyond the childhood hope that the universe was built around us. It asks us to find meaning without cosmic privilege.

That is not a loss of dignity.

It is the beginning of maturity.

A child may believe the whole household exists for him. Growing up means learning that he is loved, but not central to everything. Humanity must undergo a similar intellectual growth.

We matter to one another.

We matter to the living world.

We matter because conscious beings can love, suffer, think, create, and care.

We do not need to be the center of the universe to matter.

Meaning Without Centrality

Some people fear that if human beings are not cosmically central, then human life has no meaning.

This fear is understandable, but mistaken.

A child does not become less valuable because Earth is small.

A mother's love does not become less real because the Milky Way has hundreds of billions of stars.

Justice does not become less important because galaxies are distant.

Suffering does not become less serious because the universe is old.

Human dignity does not depend on being at the center of the cosmos.

Meaning is not a matter of location.

Meaning comes from consciousness, relationship, love, truth, compassion, creativity, responsibility, and the capacity to suffer and to reduce suffering. These things matter because they matter to conscious beings.

Religion often says that meaning must come from above. Science-based atheism says meaning arises within life itself. We do not need to be the center of the universe to care about one another. We do not need cosmic privilege to build moral lives.

In fact, cosmic humility may improve morality.

If we are not the chosen center of creation, we may become less arrogant. If Earth is our only known home, we may care for it more deeply. If human life is brief, we may value it more. If no divine parent will rescue us from our mistakes, we may become more responsible.

A universe without human centrality does not destroy meaning.

It removes flattery.

Flattery is not truth.

The Strongest Religious Reply

A thoughtful believer may say:

“The vastness of the universe does not disprove God. God could create a vast universe and still care about human beings. The size of the cosmos does not limit divine love. A parent can care about one child even in a large house. Perhaps the vastness of the universe shows God’s greatness, not human insignificance.”

This is the strongest religious reply, and it deserves fair treatment.

It is true that vastness alone does not disprove God. A believer can always say that God chose to create a vast universe. A believer can say that human beings are spiritually important even if physically small. A believer can say that God’s concern is not measured by cosmic size.

But this reply is a defense of possibility, not evidence.

It shows that belief can be preserved after modern astronomy. It does not show that modern astronomy points toward belief.

There is a difference.

When the universe was thought to be small, layered, and Earth-centered, religion used that picture as support. Earth seemed central. Heaven seemed above. Human beings seemed placed in the middle of divine attention.

When science showed that this picture was wrong, religion adapted. It said central location does not matter. It said vastness is compatible with God. It said the universe can be physically large but spiritually human-centered.

That may be possible.

But it is also a retreat.

The religious claim has moved from "the universe clearly centers on us" to "the universe does not need to center on us for God to exist."

That is a weaker claim.

Science has not proved God impossible by showing cosmic vastness. But it has weakened the old human-centered religious imagination.

Evidence-Based Response

The evidence-based response is careful.

The size and age of the universe do not logically disprove every possible God. But they strongly challenge the idea that the universe is obviously designed around human beings.

The evidence shows that Earth is one small planet. The Sun is one ordinary star. The Milky Way is one galaxy among billions. Human beings appeared very late in cosmic history. Most of the universe is hostile to life. The cosmos is not arranged in any obvious physical way around human existence.

A believer may still add God behind all this.

But God is not required to explain it.

Astronomy, physics, cosmology, chemistry, geology, and biology explain our place in the universe without divine centrality. The natural explanation is simpler, stronger, and more honest.

We are not the center.

We are not the reason galaxies exist.

We are not the purpose of cosmic expansion.

We are not the obvious goal of 13.8 billion years of cosmic history.

We are a recent form of life on a small planet in a vast universe.

This may feel humbling. But humility is not defeat. It is the beginning of truth.

Science-based atheism accepts this truth without turning the universe into a stage for human religious drama.

That is its strength.

Chapter Summary

For most of human history, people imagined the universe in human-centered ways. Earth seemed fixed. Heaven seemed above. The stars seemed like signs. Human beings seemed to stand near the center of divine concern.

Modern astronomy changed this picture completely.

Earth is a small planet orbiting an ordinary star. The Sun is one star in the Milky Way. The Milky Way is one galaxy among billions. The universe is about 13.8 billion years old, and human beings appeared very late in its history. Most of the universe is not friendly to life.

This vastness does not logically disprove every possible God. But it does weaken the old religious claim that the cosmos is obviously built around human beings.

The universe is not a human-centered stage.

It is a vast natural reality.

That reality is better explained by science than by ancient religious imagination.

Key Points

The old religious cosmos often placed human beings near the center of creation.

Modern astronomy shows that Earth is not central.

The Sun is an ordinary star, not a divine lamp made only for us.

The Milky Way is one galaxy among billions.

The observable universe may not be the total universe.

Cosmic time makes human history extremely recent.

Most of the universe is hostile to human life.

Vastness does not disprove every possible God, but it weakens human-centered religion.

Meaning does not require cosmic centrality.

Science-based atheism accepts cosmic humility without turning mystery into God.

Transition to the Next Chapter

The first lesson of modern cosmology is humility.

The universe is not arranged around us in the simple way ancient religion imagined. Earth is not the center. The Sun is not unique. The Milky Way is not the whole universe. Human beings are recent arrivals in a vast cosmic history.

But to understand how deeply this changed religion, we must go back to the older sky.

For ancient people, the sky was not space-time, radiation, galaxies, and cosmic expansion. It was heaven. It was divine territory. It was the place of gods, angels, omens, judgment, and sacred meaning.

In the next chapter, we will see how the sacred sky became the physical cosmos.

Sources

Carl Sagan, *Cosmos*

Stephen Hawking, *A Brief History of Time*

Sean Carroll, *The Big Picture*

Brian Greene, *The Fabric of the Cosmos*

Steven Weinberg, *The First Three Minutes*

Martin Rees, *Just Six Numbers*

NASA resources on the age, size, structure, and expansion of the universe

Stanford Encyclopedia of Philosophy, "Cosmology and Theology"

Chapter 2: From Sacred Sky to Physical Cosmos

When the Sky Was Heaven

Before the sky became science, it was heaven.

Ancient people did not look upward and see space-time, plasma, galaxies, redshift, radiation, or cosmic expansion. They saw power. They saw mystery. They saw danger. They saw beauty too far away to touch and too large to ignore.

The Sun rose like a ruler.

The Moon changed shape like a living sign.

The stars returned in silent patterns.

The planets wandered strangely.

Comets appeared without warning.

Eclipses darkened the day.

Thunder shook the world.

Lightning tore the sky.

Rain fell from above and made crops live.

It is not difficult to understand why human beings made the sky sacred. The sky was unreachable. It was powerful.

It controlled light, warmth, weather, seasons, crops, travel, and survival. It stood above human life in every sense: physically, emotionally, and spiritually.

So the human mind did what it often does in the presence of mystery.

It created meaning.

The sky became the home of gods. Heaven was above. Earth was below. Divine beings watched. Prayers rose upward. Blessings came down. Punishment also came down. The human world stood between hope and fear, between heaven and the grave.

For thousands of years, the sky was not simply a physical reality.

It was a religious map.

Then science changed the map.

The sky did not become less magnificent.

It became less supernatural.

Heaven slowly became cosmos.

The Old Sacred Geography

Many ancient cultures imagined reality in layers.

Heaven above.

Earth here.

The underworld below.

This structure made emotional sense because it matched ordinary experience. Light came from above. Rain came from above. The Sun and Moon were above. Birds rose upward. Smoke rose upward. The dead were buried below. Darkness belonged to caves, graves, and depths.

So "above" became sacred.

"Below" became dangerous.

The vertical structure of the world became a moral structure. Heaven above was pure, powerful, and divine. Earth was the human middle ground. The world below was shadow, death, punishment, or mystery.

This was not science.

It was symbolic geography.

But to ancient people, symbolic geography often felt physically real. Heaven was not merely an idea. It was imagined as a place. God was above the world. Divine beings lived in the high realm. The stars were not distant suns. They were lamps, spirits, jewels, signs, or sacred beings. The sky was not empty space. It was holy territory.

This sacred geography shaped prayer, ritual, architecture, language, and imagination.

People lifted their hands upward.

Temples rose toward heaven.

Mountains became sacred because they seemed closer to the gods.

Smoke from sacrifice rose into the sky.

Prayer was imagined as rising.

Blessing was imagined as descending.

Even today, the old language remains. We speak of "heaven above." We look upward when thinking of God. We say someone has "gone up" to heaven. Religious art often places the divine in the sky.

This shows how deeply the old sacred sky entered the human mind.

But modern astronomy changed the physical meaning of "above."

Above a person standing in New Jersey is not the same direction as above someone standing in India, Australia, or South America. Earth is a sphere. People stand on all sides of it. The sky is not a fixed roof over a flat world. It is space extending in every direction.

Once this is understood, the old sacred geography becomes symbolic, not literal.

Heaven can no longer be simply "up" in the old physical sense.

Science did not merely move a few stars.

It changed the structure of religious imagination.

The Sky as a Field of Signs

Before science, the sky was not only sacred.

It was interpreted.

People watched the heavens for messages.

A comet might mean disaster.

An eclipse might mean divine anger.

The movement of planets might suggest war, fertility, kingship, famine, or fortune.

The stars might guide farming, travel, ritual, prophecy, or personal destiny.

This habit was not accidental. Human beings are pattern-seeking animals. We naturally look for meaning in repeated events. The sky seemed orderly, so people assumed it carried intention. If the heavens moved with such beauty and power, perhaps they were speaking.

Astrology grew from this ancient impulse.

It connected the movement of celestial bodies with human destiny. It treated the sky as a written code, a divine or cosmic script that could be read by those who knew the signs.

This was a deeply human mistake.

The sky does have patterns.

But pattern is not the same as message.

The planets move. But they do not move to advise kings, bless marriages, warn armies, or shape the personality of a newborn child. Stars shine. But they do not shine to announce human fate. Comets appear. But they are not moral warnings. Eclipses happen. But they are not threats from heaven.

Science did not remove pattern from the sky.

It removed false meaning.

The sky became more lawful, but less personal.

More precise, but less prophetic.

More astonishing, but less flattering.

This was one of the first great losses for religious imagination.

The heavens were no longer a message system directed at human beings.

They were a physical reality.

From Sacred Lamps to Burning Stars

To ancient eyes, stars looked like points of light.

They seemed small, fixed, and decorative. They could be imagined as lamps, souls, divine signs, holes in a heavenly dome, or jewels placed in the night.

Modern astronomy revealed something far more powerful.

Stars are suns.

They are enormous spheres of hot plasma, powered by nuclear fusion. They are born from clouds of gas and dust. They shine for millions or billions of years. They change. They age. They die. Some explode. Some collapse. Some become white dwarfs, neutron stars, or black holes. Their deaths scatter heavy elements into space. Those elements later become part of new stars, planets, oceans, rocks, and living bodies.

The star is no longer a sacred lamp.

It is a natural furnace.

But what a furnace.

This is the strange beauty of science. It removes the supernatural story and gives us something greater.

A star is not less wonderful because it is natural. It is more wonderful because we now know what it is. It is not a small light hung above Earth for human use. It is a massive physical object, burning across enormous distance and time.

The old religious sky made stars emotionally meaningful by placing them near human affairs.

Science made stars magnificent by revealing their true scale.

This is one of the central lessons of this book.

Science-based atheism does not make reality smaller.

It makes reality more honest.

The stars do not need God to be beautiful.

They are beautiful because they are real.

The Telescope Broke the Sacred Ceiling

The telescope was one of the great instruments that broke the sacred sky.

When Galileo pointed his telescope upward, the heavens changed.

The Moon had mountains and valleys.

Jupiter had moons of its own.

Venus showed phases.

The Sun had spots.

The Milky Way was filled with countless stars.

The old idea of a smooth, perfect, divine heaven began to collapse. The sky was not a flawless spiritual realm. It was made of physical bodies. Those bodies had surfaces, motions, shadows, blemishes, and measurable behavior.

This mattered deeply.

If the Moon has mountains, it is not a perfect heavenly lamp.

If the Sun has spots, it is not a flawless divine disk.

If Jupiter has moons, not everything revolves around Earth.

If the Milky Way contains countless stars, the night sky is far richer than the naked eye can know.

The telescope did not simply add information.

It disciplined imagination.

It told human beings: you have been looking at the sky, but you have not really seen it.

This is what science often does.

It corrects what seems obvious.

Earth feels still, but it moves.

The Sun appears to circle us, but it does not.

The heavens look like a dome, but they are not.

The stars look small, but many are larger than the Sun.

The sky looks like a ceiling, but it is space.

The old sacred ceiling did not survive careful observation.

That is why the telescope belongs in the story of science-based atheism. It did not prove there is no God in one single stroke. But it made one thing clear: the visible sky did not need sacred geography to explain it.

The sky was physical.

Physical things can be studied without God.

Mathematics Replaced Divine Motion

After the telescope came the deeper triumph of mathematics.

Kepler showed that planets move according to mathematical patterns. Newton showed that the same law of gravitation applies to falling objects on Earth and celestial bodies in the sky. Later science refined and expanded this picture, but the basic transformation was already profound.

The planets did not need angels to push them.

The Moon did not need divine command to stay in orbit.

The heavens did not need a separate sacred physics.

Motion became natural.

This was a great turning point.

In the old religious imagination, heavenly motion was often tied to divine order. The sky moved because God, gods, angels, or celestial intelligences sustained it. But once motion could be described by natural law, the supernatural mover became unnecessary.

This does not mean every believer immediately stopped believing. Many did not. Some said God created the laws. Some said mathematics revealed divine wisdom. Some said natural law was God's method.

But the role of God had changed.

God no longer explained the movement itself.

The law explained the movement.

That is the retreat.

A believer may still place God behind the law. But the science does not need that step. The calculation works without prayer. The orbit is predicted without scripture. The eclipse comes at the expected time without divine interpretation.

Mathematics did what theology could not do.

It explained the sky.

Physical Laws Are Not Divine Commands

This point deserves special attention because religious arguments often depend on a confusion.

People hear the phrase "laws of nature" and assume that every law must have a lawgiver. Human laws need lawmakers, so physical laws must need a divine lawgiver.

But physical laws are not like traffic laws, tax laws, or criminal laws.

Human laws are commands. They are written by people, enforced by courts, and changed by governments. They tell people what they may or may not do.

Physical laws are different.

They are descriptions of how nature behaves. When scientists speak of the law of gravity, they do not mean that gravity received a command and obeyed it. They mean that gravity shows regular behavior that can be described mathematically.

A physical law is not a divine order written into the sky.

It is a human description of natural regularity.

This matters because the religious argument "laws need a lawgiver" borrows emotional force from human law and applies it wrongly to nature.

Nature does not obey laws the way citizens obey government.

Nature behaves regularly, and science describes that regularity.

Once this is understood, the need for a divine lawgiver weakens.

The planets do not move because God issued a command.

They move because physical reality has regular patterns that mathematics can describe.

That is a stronger explanation than divine decree.

Einstein and the Deeper Cosmos

Newton gave humanity a lawful universe.

Einstein made the universe stranger and deeper.

Before Einstein, space often seemed like an empty container, and time seemed like one universal clock ticking everywhere in the same way. Einstein changed that picture. Space and time were not separate, fixed backgrounds. They were connected. Gravity was not merely a force pulling across empty space. It was related to the curvature of space-time.

This was not common sense.

But science is not required to obey common sense. It is required to follow evidence.

Einstein's work showed that the universe is not the simple stage that human imagination once assumed. Space, time, gravity, matter, energy, light, and motion are woven together in ways ordinary experience does not reveal.

This matters for the God question.

Religious arguments often rely on common sense. Everything must have a cause. Time must have a beginning in the ordinary sense. Space must be a container. Nothing must mean simple emptiness. A law must have a lawgiver.

But modern physics teaches caution.

Reality at cosmic scale may not behave like the small world of human daily life. We cannot simply take the intuitions of ordinary experience and apply them to the origin of the universe.

This is one reason science-based atheism is stronger than traditional religious explanation.

It allows reality to be strange.

Religion often forces reality into human-shaped ideas: father, king, judge, maker, command, plan, purpose.

Science allows reality to correct human imagination.

Einstein's universe is not easy. But it is evidence-based. It is mathematical. It makes predictions. It can be tested. It has explained the bending of light, the behavior of gravity, the orbit of Mercury, black holes, gravitational waves, and features of modern cosmology.

God is not needed to make space-time curve.

Matter and energy do that.

From Heaven to Cosmic History

Modern cosmology changed the sky in another way.

It turned the universe into history.

In the old sacred sky, heaven often seemed eternal, fixed, and complete. God created the world, arranged the heavens, placed the stars, and established order. The sky was a divine structure.

But modern cosmology shows a universe that changes.

Stars are born and die.

Galaxies form and collide.

Planets emerge from disks of dust and gas.

Black holes form from collapsed stars.

Space expands.

Radiation cools.

The universe has a past.

This is a major shift.

The cosmos is not a static temple.

It is a natural process.

It has development, structure, transformation, and evidence of earlier states.

Light itself carries history. When we look at distant galaxies, we see them as they were long ago. The farther we look, the further back in time we see. The night sky is not only space. It is memory.

Ancient religion saw the sky as divine presence.

Modern science sees it as cosmic history.

This does not make the universe less moving.

It makes it more moving.

We are not looking at decorations above Earth. We are looking into the past of reality itself.

That is a deeper form of wonder than sacred imagination ever gave us.

The universe is not a ceiling over human life.

It is an unfolding natural history of matter, energy, space, time, stars, galaxies, and eventually life.

No divine manager is needed in that description.

The Sacred Sky Became Testable

One of the strongest differences between religion and science is testability.

A religious story may say that God placed the stars in heaven. But how do we test that?

A sacred text may say that heaven is above. But what observation could confirm it?

A myth may say that an eclipse is a divine warning. But what prediction does that make?

Science asks different questions.

How far away are the stars?

What are they made of?

How fast are galaxies moving?

How old is the light reaching us?

What happens to light as space expands?

What evidence remains from the early universe?

What does gravity do to light?

Can the model predict what we observe?

This is the transformation from sacred sky to physical cosmos.

The sky became measurable.

Once the sky became measurable, religious explanation weakened.

A measurable reality does not need myth. It needs observation, mathematics, instruments, and correction.

The telescope measured.

Spectroscopy measured.

Photography measured.

Radio astronomy measured.

Space telescopes measured.

Satellites measured.

The cosmic microwave background was measured.

Galactic redshifts were measured.

The universe became a subject of evidence.

This is why modern cosmology matters so much in the argument for science-based atheism. It does not merely say religion is wrong. It shows that the very region religion once claimed as heaven is better understood through natural explanation.

Heaven became data.

That is one of the great turning points in human thought.

From Sacred Wonder to Scientific Wonder

A common fear is that science removes wonder from the sky.

It does not.

It removes false explanation.

The Sun is not less wonderful because it is a star powered by nuclear fusion.

The Moon is not less beautiful because it is a rocky body shaped by impact, gravity, and time.

The stars are not less moving because they are distant suns.

Galaxies are not less majestic because they are systems of stars, gas, dust, dark matter, and gravity.

The cosmic microwave background is not less astonishing because it is radiation instead of revelation.

Science does not flatten the universe.

It deepens it.

Religious wonder often depends on mystery remaining unexplained. Scientific wonder grows because explanation reveals more mystery, more structure, more scale, and more truth.

This is one of the main emotional tasks of this series: to show that science-based atheism is not a dry or empty worldview. It is not merely the denial of God. It is the acceptance of reality without sacred illusion.

The sky does not need to be heaven in order to inspire awe.

It is enough that it is real.

The Strongest Religious Reply

A thoughtful believer may say:

"Science has explained the physical sky, but it has not disproved God. Heaven may not be a physical place above the clouds. God may exist beyond space and time. The stars may be natural objects, but God may still be the creator of nature."

This is a serious reply.

It is true that modern astronomy does not disprove every possible concept of God. A believer can reinterpret heaven symbolically. A believer can say God is beyond the physical universe. A believer can say that science studies creation, while theology studies the creator.

But we should notice what has happened.

Religion has moved.

The old heaven above the world has been surrendered as a literal physical claim. The stars are no longer divine lamps. Eclipses are no longer heavenly warnings. Planets are no longer spiritual messengers. The sky is no longer sacred geography in any scientific sense.

God has been moved beyond the physical evidence.

This may preserve belief, but it gives up explanation.

When believers say God is beyond space and time, they are no longer using God to explain the visible sky in the old way. They are making a more abstract claim, one that is harder to test and less connected to the physical universe we actually observe.

This is the familiar pattern.

Science explains the phenomenon.

Religion moves God elsewhere.

That movement is not victory for religion.

It is retreat.

Evidence-Based Response

The evidence-based response is clear.

The sky was once one of the strongest sources of religious imagination. It was heaven, divine territory, a field of signs, a place of gods and judgment. The Sun, Moon, planets, stars, comets, storms, and eclipses were often interpreted through religion.

Science changed this completely.

Stars are suns.

Planets are natural bodies.

Eclipses are predictable alignments.

Comets follow paths.

Galaxies are vast systems of stars.

Space is not a sacred ceiling.

Physical laws are descriptions of natural regularity, not commands from a divine lawgiver.

Gravity does not need angelic guidance.

The universe has a physical history.

None of this requires God as a scientific explanation.

A believer may still say God created the whole system. But that is not what the evidence itself shows. The evidence shows natural processes, physical law, cosmic history, and measurable reality.

The sacred sky became the physical cosmos.

This does not remove awe.

It removes illusion.

It does not make the universe smaller.

It makes it more real.

Science-based atheism stands on this honesty. It does not deny wonder. It refuses to mistake wonder for evidence of God.

Chapter Summary

For ancient people, the sky was heaven. It was divine territory, a place of signs, gods, angels, judgment, and sacred meaning. The Sun, Moon, planets, stars, comets, storms, and eclipses were often interpreted through religion.

Science changed this completely.

The telescope showed that the heavens were physical. Mathematics showed that planetary motion followed natural law. Newton united Earth and sky under the same physical principles. Einstein changed the meaning of space, time, and gravity. Modern cosmology turned the universe into a dynamic physical history.

The sky did not become less magnificent.

It became less supernatural.

The sacred sky became the physical cosmos.

God may still be believed in as a personal or philosophical idea, but God is no longer needed to explain the sky.

That is the central lesson of this chapter.

Key Points

Ancient people often imagined the sky as heaven or divine territory.

The old universe was arranged in sacred layers: heaven above, Earth here, underworld below.

The sky was interpreted as a field of signs, omens, and divine messages.

Modern astronomy showed that stars are suns and planets are natural bodies.

The telescope helped destroy the idea of perfect sacred heavens.

Mathematics and physics explained celestial motion without divine movers.

Physical laws are descriptions of nature, not commands requiring a divine lawgiver.

Einstein made space, time, and gravity part of one natural framework.

Modern cosmology made the universe a physical history, not a sacred ceiling.

Science did not remove wonder from the sky. It removed supernatural explanation.

Transition to the Next Chapter

The sacred sky became the physical cosmos, but the transformation was not complete until space and time themselves were reimagined.

For centuries, people thought of space as an empty stage and time as a universal clock. Events happened in space. Moments passed in time. God, if imagined as creator, was often thought to stand outside this stage and start the cosmic drama.

Then Einstein changed the stage itself.

In the next chapter, we will see how relativity gave the universe a new meaning of space, time, gravity, and motion.

Sources

Galileo Galilei, *Sidereus Nuncius*

Isaac Newton, *Mathematical Principles of Natural Philosophy*

Albert Einstein, *Relativity: The Special and the General Theory*

Stephen Hawking, *A Brief History of Time*

Sean Carroll, *The Big Picture*

Brian Greene, *The Fabric of the Cosmos*

Carl Sagan, *Cosmos*

John Hedley Brooke, *Science and Religion: Some Historical Perspectives*

Chapter 3: When Einstein Changed Space and Time

When the Stage Itself Changed

For most of human history, people imagined the universe as a stage.

Space was the place where things existed.

Time was the clock by which events passed.

Gravity was the force that pulled objects downward.

The Sun rose. The Moon crossed the night. Stars shone. Stones fell. Winds blew. People were born, lived, prayed, suffered, and died. Everything seemed to happen on the same great background: space beneath us, time around us, heaven above us.

This felt obvious.

Space seemed like an empty container.

Time seemed like a steady river.

The universe seemed like a vast theater in which matter moved and human life unfolded.

In religious imagination, God often stood outside this theater. God created it. God arranged it. God governed it.

God placed the lights in the sky. God watched the human drama below. God was the maker of the cosmic stage and the ruler of its events.

Then Einstein changed the stage itself.

He showed that space and time are not simple, separate, fixed backgrounds. They are woven together. They can stretch, curve, and behave in ways human common sense never imagined. Gravity is not merely a mysterious pull across emptiness. In general relativity, gravity is connected to the shape of space-time itself.

This was a stunning change.

The universe was not a simple box filled with objects.

The universe was a dynamic physical reality.

Space and time were no longer passive scenery.

They were part of nature.

That matters deeply for the question of God.

Religion often depends on human common sense. Everything must have a cause. Time must begin in an ordinary way. Space must be a container. A law must have a lawgiver. A universe must have a maker in the same way a house has a builder.

But Einstein taught humanity a difficult lesson:

The universe is not required to obey our common sense.

Reality is often stranger than the mind that first imagines it.

The Comfortable Universe Before Einstein

Before Einstein, Newtonian physics had given humanity a powerful and successful picture of nature.

Newton's laws explained motion with extraordinary accuracy. Gravity connected the fall of an apple with the motion of the Moon and planets. The same force that pulled objects toward Earth also helped hold celestial bodies in their paths. For ordinary speeds, ordinary distances, and most everyday purposes, Newton's physics still works beautifully.

But beneath that success was a picture that felt comfortable.

Space was absolute. It existed like a great invisible container. Objects moved through it.

Time was absolute. It ticked everywhere in the same way, for everyone, in every place.

Gravity was an attractive force between masses.

This picture worked so well that it seemed almost final.

It also fit easily with a certain kind of religious imagination. God could be imagined as the great designer of the cosmic machine. Space was the stage. Time was the clock. Laws were the rules. Matter was the machinery. God was the maker of the system.

This view was more sophisticated than ancient mythology. It no longer required gods to throw lightning or angels to push planets. But it still allowed many people to imagine a universe built like a grand machine by a divine engineer.

Einstein did not simply improve this picture.

He transformed it.

He showed that space and time are not the simple absolutes they appear to be. The cosmic stage is not fixed in the way common sense suggests. The clock of the universe is not one universal clock ticking identically for all observers.

This was not easy to accept.

But science does not ask whether truth is easy.

Science asks whether evidence supports it.

Special Relativity in Plain Language

Einstein's special theory of relativity began with a simple but profound problem.

Light behaves in a strange way.

No matter how fast an observer moves, the speed of light in a vacuum remains the same. This seems impossible to ordinary common sense. In daily life, speeds add and subtract. If you walk forward inside a moving train, your speed relative to the ground is the speed of the train plus your walking speed.

That is common sense.

But light did not fit ordinary intuition.

Einstein accepted the evidence and allowed common sense to change.

The result was special relativity.

Special relativity showed that time and space are linked to motion. Time does not pass in exactly the same way for all observers. Length and time can depend on relative motion. Nothing with mass can travel faster than light in vacuum. Matter and energy are deeply connected.

This is where the famous equation comes from:

$E = mc^2$

That equation is not only a scientific formula. It is a window into deeper reality. Matter and energy are not completely separate. They are related. The universe is more unified than ordinary experience suggests.

This matters because religious thinking often assumes that human intuition is a trustworthy guide to ultimate reality.

But special relativity shows that intuition can fail.

Time feels absolute.

It is not.

Space feels fixed.

It is not so simple.

Matter feels completely different from energy.

It is not.

The universe does not bend itself to human expectation.

Human expectation must bend to the universe.

That is the scientific attitude.

And that attitude repeatedly weakens God-based explanations.

General Relativity: Gravity Becomes Geometry

Einstein's general theory of relativity went further.

It changed the meaning of gravity.

In Newton's picture, gravity was a force of attraction between masses. Earth attracts the Moon. The Sun attracts Earth. Objects fall because Earth pulls them.

This was powerful and useful, but it left deep questions.

How does gravity act across empty space?

Why does it work the way it does?

Einstein offered a new picture.

Mass and energy shape space-time. Objects move along paths determined by that curved space-time. What we experience as gravity is related to the geometry of space-time itself.

A simple image may help, though every image is imperfect. Imagine a stretched fabric. Place a heavy ball on it, and the fabric curves. Smaller objects moving nearby follow paths shaped by that curve. The real universe is not a flat fabric, but the image gives a hint: matter changes the structure around it, and that structure guides motion.

This is extraordinary.

Gravity is no longer just a force acting on a fixed stage.

The stage itself participates.

Space-time is dynamic.

This was a deeper natural explanation than before. It did not invoke divine control. It did not say God tells planets how to move. It did not say the Sun holds Earth by spiritual command. It described a physical relationship between mass, energy, space, time, and motion.

That is why general relativity matters in this book.

It moved the universe further away from divine management and deeper into natural explanation.

God was not needed to make gravity work.

Physics was.

Light Bends, Evidence Speaks

A beautiful theory is not enough.

Science demands evidence.

One of the great tests of general relativity involved light passing near the Sun. Einstein's theory predicted that gravity would bend light. During a solar eclipse in 1919, astronomers observed stars near the Sun's apparent position and found results consistent with Einstein's prediction.

The world listened.

Light had bent.

This was not merely a technical triumph. It was a demonstration of how science earns authority. Einstein did not win because his idea was comforting. He did not win because he belonged to a priesthood. He did not win because his theory was written in sacred language.

His theory had to face reality.

Would light behave as predicted?

Observation answered.

This is a crucial difference between science and religion.

A religious claim may be protected by tradition. A doctrine may survive because it is sacred. A scripture may be defended through reinterpretation. But a scientific theory must face the world.

If reality says no, the theory must change.

If reality says yes, the theory grows stronger.

Relativity became powerful because it explained, predicted, and survived testing.

God did not bend light in a way that theology could calculate.

General relativity did.

That is why natural explanation becomes stronger than supernatural explanation. It does not merely declare. It demonstrates.

Common Sense Is Not the Final Judge

Einstein's work teaches one of the most important lessons of modern thought:

Common sense is limited.

This is not an insult to common sense. We need common sense for daily life. It helps us cross roads, cook food, use tools, avoid danger, and communicate with other people. It works well in the human-sized world.

But the universe is not limited to the human-sized world.

At very high speeds, common sense fails.

Near massive objects, common sense fails.

At cosmic scales, common sense fails.

At quantum scales, common sense fails.

This matters because many arguments for God rely on common sense stretched far beyond its proper place.

Everything that begins must have a cause.

Nothing can come from nothing.

Every law must have a lawgiver.

Every design must have a designer.

Every order must have an ordering mind.

These sentences sound persuasive because they fit ordinary life. Houses need builders. Books need writers. Machines need engineers. Human laws need lawmakers.

But the universe is not a house.

Physical laws are not human laws.

The beginning of space-time is not like the beginning of a building.

The origin of the universe is not an event inside the universe in the same ordinary way that lighting a candle is an event inside a room.

Einstein does not answer every question about cosmic origin. But relativity warns us against simple religious answers based on ordinary intuition.

Reality is deeper than common sense.

That is why science must lead.

God Outside Time?

Some believers respond to relativity by saying that God exists outside time.

This idea is common in modern theology. If time is part of the universe, then perhaps God is timeless. God does not exist before the universe in a normal temporal sense. God is beyond time, beyond space, beyond matter, and beyond physical law.

This sounds sophisticated.

It may be philosophically interesting.

But we must ask a serious question:

What does it explain?

If God is outside time, how does God act?

If God is outside space, where is God?

If God is immaterial, how does God create matter?

If God is timeless, how does God decide, choose, intend, or begin creation?

If God has no before and after, what does it mean for God to cause anything?

These questions are not small.

The claim that God is beyond space and time may protect God from scientific investigation, but it also removes God from clear explanation.

A timeless, spaceless, immaterial, invisible mind is not an explanation that science can test, measure, model, or use.

It is a theological assertion.

Some believers may accept it by faith. But faith is not evidence.

Science-based atheism asks for more than a protected mystery. It asks whether the claim explains reality better than natural law.

In cosmology, it does not.

Relativity gives us a mathematical framework that explains gravity, motion, space-time, and cosmic structure. "God outside time" gives us no comparable framework.

It does not predict the bending of light.

It does not calculate the orbit of Mercury.

It does not explain black holes.

It does not model cosmic expansion.

It does not give a testable account of the universe.

It preserves belief, but it does not improve explanation.

Einstein's God Was Not the God of Religion

Many people quote Einstein as if he supported traditional belief in God.

This is misleading.

Einstein sometimes used the word God. He spoke of cosmic religious feeling. He admired the rational order of nature. He resisted crude materialism and narrow thinking. He felt deep awe before the universe.

But Einstein did not believe in a personal God who listens to prayers, intervenes in history, rewards believers, punishes unbelievers, suspends natural law, or judges human beings after death.

Einstein's "God" was closer to awe at the order and mystery of nature than to the God of ordinary religion.

This distinction matters.

Religious apologists often borrow Einstein's language but ignore his meaning. They hear the word God and assume he meant their God. But Einstein's cosmic language does not support the personal creator of popular religion.

Einstein's work actually strengthened the natural understanding of the universe.

He did not explain gravity by divine intention.

He did not explain space-time by scripture.

He did not explain cosmic order by prayer.

He gave mathematical physics.

That is the point.

A scientist may feel awe and still not need God.

A person may use religious language poetically and still reject supernatural religion.

Einstein's importance in this book is not that he was an atheist in a simple way. His views were more complex. His importance is that his science made the universe more deeply natural.

Space, time, and gravity became part of physical explanation.

God became unnecessary for that explanation.

The Universe Became Dynamic

Einstein's equations also changed cosmology.

They did not merely describe objects moving through space. They had implications for the structure of the universe itself. The universe could be dynamic. It could expand or contract. Space-time itself could have cosmic history.

Einstein initially favored a static universe, as many scientists did at the time. He introduced a cosmological constant into his equations partly to allow a static cosmic model. Later, observations of cosmic expansion moved science in another direction.

This is important.

Einstein was brilliant, but not infallible.

Science did not freeze his preference into doctrine. Evidence corrected the picture.

That is one of the moral strengths of science. No mind, however great, is sacred. No scientist becomes scripture. Even Einstein had to face observation.

The later discovery of cosmic expansion showed that the universe is not an eternal, unchanging stage in the old sense. It has history. It changes. It evolves. Space itself expands.

This again weakens old religious imagery.

The universe is not a finished temple placed by God in static order.

It is a dynamic physical reality.

It has development.

It has evidence.

It has structure.

It has change.

It can be studied.

And what can be studied naturally does not need divine management as its explanation.

Black Holes and the Strange Natural Universe

Relativity also opened the door to some of the strangest objects in modern science: black holes.

A black hole is a region where gravity is so strong that, within a certain boundary, not even light can escape. Such objects sound almost mythological, but they arise from physical theory and are supported by evidence.

Black holes teach another lesson.

The universe is more astonishing than religious imagination.

Ancient stories imagined heavens, underworlds, demons, divine chariots, sacred mountains, and cosmic waters. But modern physics revealed space-time curvature, gravitational waves, black holes, neutron stars, expanding space, and cosmic background radiation.

Reality is not less dramatic than religion.

It is more dramatic.

But it is dramatic without needing gods.

This is important for readers who fear that atheism makes the universe dull. A natural universe is not dull. It is overwhelming.

The difference is that science does not explain strangeness by inventing supernatural beings. It studies the strangeness itself.

Black holes are not mouths of hell.

They are not divine prisons.

They are not mythic abysses.

They are physical consequences of gravity, mass, and space-time.

Science replaces supernatural fantasy with deeper reality.

That is not a loss.

It is an intellectual victory.

Relativity and the Creator Argument

Relativity matters to the creator argument in several ways.

First, it shows that space and time are part of the physical universe, not simple background containers. This makes the question of creation more subtle. If time itself is part of the universe, then asking what happened before the universe may not be like asking what happened before breakfast.

Second, relativity shows that human intuition is not reliable at cosmic scales. This weakens simplistic arguments based on ordinary cause and effect.

Third, relativity allows the universe itself to be studied through mathematical law. The cosmos is not beyond science. It is a subject of science.

Fourth, relativity helped open the way to modern cosmology, including expanding universe models and Big Bang cosmology.

For religion, this creates a problem.

If the universe can be studied through physics, then God is not needed as the first explanatory step. God may be added later as a philosophical belief, but the scientific work proceeds without God.

This is exactly the pattern of this series.

God once explained what people did not understand.

Then science explained it better.

In the case of relativity, God is not needed to explain gravity, space-time, light bending, black holes, gravitational waves, or the large-scale structure of the universe.

Physics does that work.

The Strongest Religious Reply

A thoughtful believer may say:

"Einstein did not disprove God. Relativity explains how gravity, space, and time work, but it does not explain why there is a universe, why there are laws of physics, or why mathematics describes reality. God may still be the creator of the system Einstein described."

This is a serious reply.

It is true that relativity does not answer every ultimate question. It does not explain why there is something rather than nothing. It does not by itself prove atheism. It does not remove every philosophical mystery.

But it does show something important.

God is not needed to explain the physical behavior that relativity explains.

The bending of light does not require divine will.

The orbit of Mercury does not require scriptural interpretation.

The structure of space-time does not require theology.

Black holes do not require supernatural explanation.

The large-scale universe can be modeled without invoking God.

The believer may still say that God created the laws. But that is a different claim. It is not part of the scientific explanation. It gives no equation, no testable prediction, no mechanism, no measurement.

It is a belief placed behind the physics.

That belief may comfort some people. But it does not improve the explanation.

Evidence-Based Response

The evidence-based response is clear.

Einstein changed the meaning of space, time, gravity, and motion. His theories showed that the universe is deeper and stranger than ordinary common sense. Relativity has been tested repeatedly and has become one of the foundations of modern physics and cosmology.

This does not prove that no God can exist in any possible sense.

But it does make God unnecessary for explaining what relativity explains.

Science-based atheism does not need to claim more than that. Its strength is not overstatement. Its strength is precision.

Where physics explains, God is not needed.

Where evidence speaks, scripture is not required.

Where mathematics predicts, divine intention adds nothing.

Relativity shows that the universe is lawful, physical, and intelligible without supernatural management.

That is the central point.

The universe is not a stage on which God merely places events.

Space and time themselves are part of the natural world.

Once even the stage becomes natural, the old divine theater loses its foundation.

Chapter Summary

Before Einstein, many people imagined space as an empty container and time as a universal clock. This picture fit common sense and allowed the universe to be imagined as a fixed stage created and governed by God.

Einstein changed that picture.

Special relativity showed that time and space are linked to motion and that matter and energy are deeply related. General relativity showed that gravity is connected to the curvature of space-time caused by mass and energy. The universe became more dynamic, physical, mathematical, and strange than common sense had imagined.

Relativity does not answer every ultimate question. It does not disprove every possible God. But it makes God unnecessary as an explanation for gravity, space-time, light bending, black holes, and much of modern cosmology.

Einstein's universe is not a sacred stage.

It is a natural reality.

Key Points

Before Einstein, space and time seemed absolute and separate.

Special relativity showed that space and time are linked to motion.

General relativity reimagined gravity as connected to the curvature of space-time.

Einstein's work showed that common sense often fails at cosmic scales.

Relativity was powerful because it made testable predictions.

Einstein's use of the word God should not be confused with belief in a personal religious God.

Relativity helped make the universe a subject of scientific cosmology.

God is not needed to explain what relativity explains.

Transition to the Next Chapter

Einstein changed the stage.

Space and time were no longer simple backgrounds. Gravity was no longer merely a mysterious pull. The universe became dynamic, mathematical, and physically deeper than ordinary human imagination.

But another discovery was needed to change the cosmic story completely.

If Einstein changed the meaning of the stage, Hubble showed that the stage itself was expanding.

The Milky Way was not the whole universe. Galaxies lay far beyond it. Their light carried a message: the universe was not static.

In the next chapter, we will see how Hubble and the expanding universe changed the question of creation forever.

Sources

Albert Einstein, *Relativity: The Special and the General Theory*

Arthur Eddington, *Space, Time and Gravitation*

Stephen Hawking, *A Brief History of Time*

Sean Carroll, *From Eternity to Here*

Brian Greene, *The Fabric of the Cosmos*

Kip Thorne, *Black Holes and Time Warps*

Steven Weinberg, *The First Three Minutes*

Stanford Encyclopedia of Philosophy, “Space and Time: Inertial Frames”

Chapter 4: Hubble and the Expanding Universe

The Galaxies Were Not Standing Still

Imagine discovering that the universe is not still.

The stars do not simply hang above us.

The galaxies are not fixed forever in sacred order.

The cosmos is not a completed temple, silent and unmoving.

It is expanding.

On the largest scales, distant galaxies are moving away from us. More precisely, space itself is expanding. The farther away a galaxy is, the faster it appears to recede.

This was not how the universe appeared to ancient human beings.

To ancient eyes, the sky seemed permanent. The Sun rose and set. The Moon changed shape and returned. The stars came back season after season. Human beings were born and died. Kings rose and fell. Cities were built, burned, and rebuilt. But the heavens seemed stable.

Religion naturally built itself around that feeling.

The heavens were fixed.

The divine order was above.

God's creation was arranged.

The universe seemed like a completed structure.

Earth was the human stage beneath the sacred sky.

Then modern astronomy changed the story.

The universe was not a fixed divine building.

It was dynamic.

It had a history.

It was expanding.

This discovery changed the question of creation. If the universe is expanding now, then the universe was different in the past. If galaxies are moving apart, then long ago the observable universe was hotter, denser, and more compressed.

The universe is not merely a place.

It is a process.

That is the great lesson of Hubble.

The cosmos is not a still stage created once and left unchanged.

The cosmos is a physical history.

And history can be studied.

The Comfort of a Static Heaven

The older religious imagination often pictured the heavens as stable and complete.

God created the world.

God placed the lights in the sky.

God arranged the order of things.

God ruled over creation.

The heavens displayed divine glory.

This picture had emotional power. A fixed heaven felt secure. It gave people a sense of order. It made the universe feel like architecture: designed, arranged, completed, and watched over.

For a long time, even many scientists imagined the universe as static on the largest scale. Stars might move locally, but the cosmos as a whole was often pictured as eternal or unchanging in its broad structure.

This was comfortable.

A static universe feels like a building.

A dynamic universe feels like history.

A building invites the question, “Who built it?”

History invites a different question: “How did it develop?”

Religion often prefers the first question because it points toward a builder. Science follows the second because it points toward process, evidence, and change.

Cosmic expansion changed the emotional picture of the universe.

The universe was not a finished palace.

It was an unfolding physical reality.

This does not disprove every possible God. But it does move the discussion away from sacred architecture and toward natural history. The universe becomes something whose past can be investigated through light, distance, motion, temperature, radiation, and mathematics.

The static heaven gave religion a stage.

The expanding universe gave science a story.

The Milky Way Was Not the Whole Universe

Before Hubble, one of the great questions in astronomy concerned the nature of spiral nebulae.

Some astronomers thought these faint spiral objects were clouds of gas inside our own Milky Way. Others suspected they were "island universes," vast systems of stars far beyond our galaxy.

This was not a small disagreement.

If the spiral nebulae were inside the Milky Way, then our galaxy might be nearly the whole visible universe. But if they were separate galaxies, then the cosmos was far larger than most people had imagined.

Edwin Hubble helped settle this question.

By studying variable stars, especially Cepheid variables, Hubble estimated the distance to the Andromeda nebula. The result was astonishing. Andromeda was far outside the Milky Way. It was not a nearby cloud inside our galaxy. It was another galaxy.

The universe became larger in the human mind.

The Milky Way was not the universe.

It was one galaxy among many.

This discovery matters for the deeper argument of this book. Human beings had already lost the centrality of Earth. They had lost the perfection of the heavens. They had lost the idea that the Sun was unique. Now even our galaxy lost its privileged status.

The universe kept pushing us away from centrality.

Religion could adapt, as it often does. A believer could say that God created many galaxies. That is possible as belief. But the old emotional structure was weakened again.

The universe no longer looked like a small sacred home built around us.

It looked like an immense natural reality, mostly unknown to ancient religion.

Redshift: Light Carrying Cosmic Motion

Hubble's next great contribution involved galaxy redshifts.

Light can tell us whether a distant object is moving toward us or away from us. When light from a galaxy is shifted toward the red end of the spectrum, it suggests that the galaxy is receding from us. In modern cosmology, this is understood in relation to the expansion of space itself on very large scales.

This idea is difficult because ordinary experience does not prepare us for it.

We are used to objects moving through space. A car moves down a road. A bird moves through air. A stone flies through the sky.

Cosmic expansion is different.

The space between distant galaxies is stretching.

A common image is raisin bread rising in an oven. As the dough expands, the raisins move farther apart. No raisin needs to be at the center of expansion. From each raisin, the other raisins appear to move away. The image is imperfect, but it helps.

In the expanding universe, galaxies are like the raisins, and space is like the dough.

This means we are not at the center of expansion in the old privileged sense. Distant galaxies appear to recede from us because space is expanding everywhere on large scales. Observers in other galaxies would also see distant galaxies receding from them.

This is another correction of human pride.

At first glance, everything seems to move away from us. A religious imagination might be tempted to say, “We are central after all.” But science says no. The appearance is a result of cosmic expansion, not Earth’s special location.

Once again, science corrects the flattering interpretation.

The universe is not running away from Earth because Earth is sacred.

The universe is expanding.

Hubble’s Law and the Birth of Cosmic History

Hubble found a relationship between distance and recession speed.

In simple language, the farther away a galaxy is, the faster it appears to be moving away from us.

This relationship became known as Hubble’s law.

The importance of this discovery cannot be overstated. If distant galaxies are receding, then the universe is not static. If we mentally run the expansion backward, the observable universe was once denser and hotter.

This opened the door to Big Bang cosmology.

The phrase Big Bang can be misleading, and the next chapter will explain that carefully. But for now, the key point is this: Hubble's work helped show that the universe has a physical past.

The universe became historical.

This is a major change from the old religious picture. Sacred stories often describe creation as a divine act in the past, but they do not offer a scientific history. They do not measure redshift. They do not calculate expansion. They do not use variable stars to estimate galactic distances. They do not predict cosmic background radiation. They do not model the growth of galaxies.

Modern cosmology does those things.

The universe's past is not known through revelation.

It is studied through evidence.

This is why science-based atheism becomes powerful. It does not simply say, "There is no creator." It says, "Look at how much reality can be explained without one."

Cosmic expansion is a physical phenomenon.

It does not require divine command.

Expansion Is Not an Explosion Into Empty Space

Many people misunderstand cosmic expansion.

They imagine a huge explosion happening at one place in empty space, throwing matter outward like fragments from a bomb. In that picture, the Big Bang happens somewhere, and galaxies fly away from that point.

This is not the correct picture.

The Big Bang was not an explosion into empty space.

Space itself expanded.

There was no ordinary empty room waiting for the universe to explode into. Space is part of the universe. The expansion is not matter flying outward from a central point into preexisting emptiness. It is the stretching of space itself.

This distinction matters because religious arguments often use a simplified and misleading image.

If the Big Bang is imagined like an explosion, people naturally ask, "Who lit the fuse?" But if space-time itself is part of the physical reality being described, the question becomes much more subtle.

"Who lit the fuse?" is a human-sized question.

Cosmic origin is not a human-sized event.

This does not mean there are no deep questions. There are. It does not mean physics has explained everything. It has not. But it does mean that simple religious images often distort the science.

The expanding universe is stranger than a bomb.

It is not like a firework in the sky.

It is the expansion of the cosmic stage itself.

That is why we must be careful before turning the Big Bang into theology.

A misunderstood universe can easily become a false argument for God.

The Religious Excitement Over a Beginning

The discovery of cosmic expansion created excitement among many religious thinkers.

If the universe is expanding, then perhaps it had a beginning.

If it had a beginning, perhaps it had a beginner.

If it had a beginner, perhaps that beginner is God.

This chain of thought has emotional force.

For many believers, Big Bang cosmology seemed to support the old doctrine of creation. Some saw it as science confirming scripture. Some argued that physics had finally admitted what religion had always known: the universe began.

But this argument moves too quickly.

A scientific model of cosmic expansion is not the same as a religious creation story.

The universe being different in the past does not prove divine creation.

An early hot, dense state does not identify a creator.

A beginning, if there is a beginning in the ordinary sense, does not automatically become the God of religion.

This is one of the central lessons of this book.

Religious arguments often take a scientific discovery and add theology to it. They take expansion and add creation. They take beginning and add God. They take mystery and add divine intention.

But the added part is not science.

It is interpretation.

Science has evidence for expansion. It has evidence for an early hot, dense universe. It has evidence for cosmic microwave background radiation, light element abundance, and large-scale structure.

It does not have evidence that a personal God commanded the universe into existence.

That leap is not justified.

Lemaître and the Complexity of Science and Faith

The history of Big Bang cosmology includes an important figure: Georges Lemaître.

Lemaître was a Belgian physicist and Catholic priest. He proposed an expanding universe model and what he called the primeval atom, an early idea related to what later became Big Bang cosmology.

His life complicates simplistic stories.

Science and religion have not always been enemies in a crude sense. A priest could contribute to modern cosmology. A believer could do serious physics. Religious identity did not prevent Lemaître from thinking scientifically.

This should be acknowledged honestly.

But Lemaître's scientific work did not prove Catholicism. It did not prove God. It did not prove scripture. It did not show that the universe requires a creator.

In fact, Lemaître himself was careful not to confuse physics with theology. He understood that a scientific model should not be turned too quickly into religious proof.

This is an important lesson.

The issue is not whether a scientist is religious.

The issue is whether God is needed in the explanation.

Lemaître may have been a priest, but the expanding universe model belongs to physics, not theology.

This pattern appears throughout the history of science. Many scientists had religious beliefs. But their scientific explanations often worked without God.

The personal belief of the scientist and the explanatory power of the science are not the same thing.

The Universe Became Older Than Scripture

Cosmic expansion also helped scientists estimate the age of the universe.

Modern measurements place the age of the universe at about 13.8 billion years.

This number is difficult to feel.

Human beings live for decades.

Civilizations last for centuries or millennia.

Written history is only several thousand years old.

The universe is billions of years old.

This does not merely stretch time. It changes the meaning of human existence. The universe existed long before Earth. Earth existed long before human beings.

Stars formed and died before our solar system was born. Galaxies evolved across immense spans of time.

Ancient religious stories were created by human beings who knew almost none of this. They looked at the sky and built meaning from what they could see. That was natural. But their imagination was limited by their knowledge.

Modern cosmology has shown that the universe is not built on the scale of human stories.

It is older, larger, and stranger.

This does not make human beings worthless. It makes our appearance more remarkable. We are latecomers in a vast cosmic process. Matter became stars. Stars forged elements. Elements became planets. On one planet, chemistry eventually became life. Life became consciousness. Consciousness looked back at the universe and asked where it came from.

That is magnificent.

It does not need God to be magnificent.

It is magnificent because it is true.

A Dynamic Universe Without Divine Management

An expanding universe is a universe of process.

It has no need for constant divine adjustment in the scientific model. Galaxies move according to physical law.

Space expands according to cosmological dynamics. Gravity shapes structure. Matter and radiation evolve. Stars form. Galaxies cluster. The universe develops.

This does not mean every detail is fully understood.

It is not.

Dark matter remains mysterious.

Dark energy remains mysterious.

The exact nature of cosmic inflation remains under investigation.

Different measurements of the expansion rate of the universe have produced tensions that scientists continue to study.

But these mysteries do not automatically support God.

They support more research.

This is the difference between science and religion.

Religion often says: here is a mystery, therefore God.

Science says: here is a mystery, therefore investigate.

The expanding universe is full of unanswered questions. But unanswered questions are not evidence for divine action. They are invitations to deeper science.

The God of the gaps has always lived in the unknown.

But the unknown is not safe territory for God.

History shows that gaps close.

When they close, God retreats again.

The Emotional Shock of Cosmic Expansion

There is an emotional side to Hubble's discovery.

A static universe can feel comforting. It feels stable, finished, and permanent. It seems like a great structure, perhaps built by a great builder. It offers a sense of order.

An expanding universe feels different.

It is restless.

It is changing.

It has a past and a possible future.

It raises questions about beginning, direction, fate, and ultimate meaning.

Will the universe expand forever?

Will stars eventually burn out?

Will galaxies drift beyond one another's view?

Will the universe grow cold and dark?

These questions can be emotionally unsettling.

Religion often gives emotionally satisfying answers. God created. God sustains. God has a plan. God will bring final meaning. The universe has purpose because God gave it purpose.

Science gives a different kind of answer.

It does not promise comfort.

It follows evidence.

This is why some people resist a fully natural universe. They do not merely want to know how the universe behaves. They want the universe to care. They want the universe to have intention. They want the cosmos to be a home, not only a physical reality.

But the universe is not obligated to satisfy human emotional need.

Science-based atheism begins with that honesty.

The universe may not care about us.

But we can care about one another.

That is enough for meaning.

A Beginning Is Not a God

The strongest religious use of Hubble's discovery is the claim that expansion points to a beginning, and a beginning points to God.

This must be answered clearly.

Even if the universe had a beginning, that does not prove God.

A beginning is not a person.

A beginning is not a mind.

A beginning is not love.

A beginning is not moral perfection.

A beginning is not scripture.

A beginning is not heaven.

A beginning is not prayer.

A beginning is not the God of any religion.

At most, a beginning raises a question.

What explains it?

That question is serious. But God is only one proposed answer, and it comes with many problems of its own. If God explains the universe, what explains God? If God is uncaused, why cannot reality itself be uncaused? If God is timeless, how does God act? If God is immaterial, how does God create matter? If God is beyond space and time, what does it mean to say God caused space and time?

These are not minor difficulties.

They show that God is not a simple answer.

God is a larger mystery placed behind the first mystery.

Science may not yet have the final answer to cosmic origin. But "not yet" is not the same as "therefore God."

That is the discipline this book insists on.

The Strongest Religious Reply

A thoughtful believer may say:

"Hubble's expanding universe supports belief in God. If the universe is expanding, then it had a beginning. If it had a beginning, then it needs a cause. That cause must be beyond space, time, matter, and energy. This sounds very much like God. Therefore modern cosmology supports the creator."

This reply is serious and should not be dismissed casually.

It correctly recognizes that the expanding universe changed the debate. It also correctly notes that the origin of the universe raises deep questions.

But the reply goes beyond the evidence.

Cosmic expansion supports the idea that the observable universe evolved from an earlier hot, dense state. It does not identify a divine cause. It does not prove that the cause, if there is one, is personal, conscious, loving, moral, prayer-answering, or connected to any scripture.

It also assumes that ordinary causation applies to the universe as a whole. That may not be true in any simple way, especially if time itself is part of the universe.

Finally, even if one accepts a cause beyond the universe, calling that cause God adds many claims not contained in the evidence.

The expanding universe is science.

The creator is theology.

We should not confuse the two.

Evidence-Based Response

The evidence-based response is this:

Hubble's work helped reveal a dynamic, expanding universe. It showed that the Milky Way is not the whole cosmos and that distant galaxies are generally receding from us. This led to the modern understanding that the observable universe has a physical history.

That is a major scientific discovery.

But it does not prove God.

It does not prove creation from nothing.

It does not prove a personal creator.

It does not prove divine intention.

It does not prove that ancient scriptures understood modern cosmology.

It does not show that the universe was made for human beings.

What it shows is more precise and more powerful:

The universe changes.

The universe has a past.

The universe can be studied through evidence.

Its history can be reconstructed through light, distance, motion, radiation, and physical law.

God is not needed in that scientific explanation.

Science-based atheism does not deny the wonder of the expanding universe. It accepts the wonder and refuses to turn it into theology without evidence.

That is its strength.

Chapter Summary

Hubble helped change the human understanding of the universe.

He showed that the Milky Way is not the whole universe. Other galaxies exist far beyond it. He also helped establish the relationship between a galaxy's distance and its redshift, leading to the understanding that the universe is expanding.

This discovery transformed cosmology.

The universe was no longer a fixed sacred structure or a static cosmic stage. It became dynamic. It had a physical history. It could be studied through evidence.

Religious thinkers often interpret cosmic expansion as proof of creation. But this is a leap beyond the evidence. Expansion points to an earlier hot, dense state. It does not prove a personal God.

The expanding universe is magnificent.

But it does not require a creator.

Key Points

The old religious imagination often pictured the heavens as fixed, sacred, and complete.

Hubble helped show that the Milky Way is only one galaxy among many.

Galaxy redshift helped reveal that the universe is expanding.

Cosmic expansion means space itself expands on large scales.

The Big Bang is not an explosion into empty space.

Expansion gives the universe a physical history.

A beginning, even if accepted, is not automatically God.

The expanding universe supports modern cosmology, not religious creation claims.

God is not needed to explain cosmic expansion.

Transition to the Next Chapter

Hubble showed that the universe is expanding.

That discovery changed everything.

If the universe is expanding now, then in the past it must have been denser, hotter, and very different from the universe we see today. This idea led to what is now called Big Bang cosmology.

But the Big Bang is often misunderstood.

It is not God lighting a cosmic fuse.

It is not an explosion into empty space.

It is not a scientific version of Genesis.

In the next chapter, we will examine what the Big Bang actually explains, what it does not explain, and why it does not prove God.

Sources

Edwin Hubble, "A Relation Between Distance and Radial Velocity Among Extra-Galactic Nebulae"

Georges Lemaître, writings on the expanding universe and primeval atom

Alexander Friedmann, writings on dynamic cosmological models

Stephen Hawking, *A Brief History of Time*

Steven Weinberg, *The First Three Minutes*

Sean Carroll, *From Eternity to Here*

Brian Greene, *The Fabric of the Cosmos*

NASA resources on galaxies, redshift, and cosmic expansion

Chapter 5: The Big Bang: What It Explains and What It Does Not Prove

Not God Lighting a Cosmic Fuse

Few scientific phrases have been more misunderstood than the Big Bang.

The name itself almost invites the wrong picture.

Many people imagine empty darkness. Then God speaks. A cosmic explosion bursts outward. Matter flies in all directions. Stars, galaxies, planets, and human beings eventually appear. The ancient creation story seems to return, now dressed in scientific language.

This picture is powerful.

It is also wrong.

The Big Bang was not God lighting a cosmic fuse.

It was not an explosion inside empty space.

It was not a fireball thrown into a preexisting room.

It was not the first sentence of Genesis translated into physics.

It was not a scientific proof of any scripture.

The Big Bang is a scientific model. It describes the observable universe expanding and cooling from an earlier hot, dense state. It is based on evidence, not revelation. It belongs to physics, astronomy, and cosmology, not theology.

This distinction is essential.

The Big Bang does not say, "God created the universe."

It says that the universe we observe today was once much hotter, denser, and more compressed, and that space itself has been expanding over cosmic time.

That is a profound scientific discovery.

But it is not a religious conclusion.

Religious apologists often place God inside the mystery of cosmic beginnings. They say: the universe began, therefore God began it. But this moves too quickly. It takes a scientific model and adds theology to it. It turns a question into a doctrine. It changes "we have evidence of an early hot, dense universe" into "God did it."

That leap is not justified.

The Big Bang is not proof of God.

It is evidence that the observable universe has a physical history.

What the Big Bang Actually Means

The phrase Big Bang can mislead because a bang happens somewhere.

A bomb explodes in a room.

A firework explodes in the sky.

Material flies outward from a central point into surrounding space.

The Big Bang was not like that.

There was no empty space outside the universe waiting for matter to enter. Space itself is part of the universe. The Big Bang model describes the expansion of space, not matter flying through space from one central location.

Everywhere in the observable universe was once hotter and denser.

As space expanded, the universe cooled.

Particles formed.

Atoms formed.

Light began to travel freely.

Gravity slowly helped matter gather into stars, galaxies, and large-scale structures.

Later, stars created heavier elements.

Some of those elements became planets.

On one small planet, chemistry eventually became life.

This is not mythology. It is a scientific reconstruction based on evidence.

The Big Bang does not describe every detail of the ultimate origin of reality. It does not answer every question about why there is something rather than nothing. It does not yet give a final theory of quantum gravity. It does not tell us, with certainty, what happened at the earliest boundary where our current theories become incomplete.

But it explains a great deal.

It explains why distant galaxies are generally receding from us.

It explains why the universe contains a faint background radiation.

It explains why the universe contains certain proportions of light elements.

It explains why the universe has evolved from a hotter, denser past to its present structure.

That is already remarkable.

The Big Bang is not a complete answer to every cosmic question.

But it is far better than a creation story.

Creation Story Versus Scientific Model

Religious creation stories usually begin with divine intention.

God decides.

God commands.

God speaks.

God creates heaven and Earth.

In these stories, the universe begins as an act of will.

Big Bang cosmology is different.

It begins with evidence.

Light from galaxies is redshifted.

The universe is expanding.

There is cosmic microwave background radiation.

There are patterns in the distribution of galaxies.

There are measurable abundances of hydrogen, helium, and other light elements.

There are mathematical models that can be compared with observation.

This is not a sacred narrative.

It is a scientific explanation.

Creation stories usually explain the universe in human terms. A mind intends. A voice commands. A maker makes. A plan unfolds. These ideas feel natural because human beings understand human action. We know what it means to build a house, write a book, make a tool, or choose a purpose.

So we project this human pattern onto the universe.

But the universe is not a house.

It is not a clay pot.

It is not a temple.

It is not a book.

It is not a machine built in a workshop.

It is reality itself, including space, time, matter, energy, and the laws that describe their behavior.

To say "God made the universe" may feel satisfying because it gives the mind a familiar pattern: maker and made. But familiar is not the same as true.

The Big Bang does not give us a maker.

It gives us expansion, temperature, density, radiation, matter, and cosmic evolution.

The scientific story is less human-shaped.

That is why it is more trustworthy.

Why the Big Bang Felt Religious to Many People

It is easy to understand why many believers welcomed the Big Bang.

For a long time, some scientists believed in a static or eternal universe. If the universe had always existed in roughly the same general state, it was harder to use a beginning as evidence for God.

But an expanding universe seemed different.

If the universe expanded from an earlier hot, dense state, then perhaps it had a beginning.

If it had a beginning, perhaps it needed a beginner.

This seemed to fit religious creation.

Some believers felt that science had finally confirmed what scripture had said all along: the universe began.

But this interpretation has serious problems.

First, the Big Bang is not the same as creation from absolute nothing.

Second, the Big Bang does not identify a creator.

Third, different religions have different creation stories, and Big Bang cosmology does not confirm any one of them.

Fourth, a universe with a hot, dense early state does not prove a personal God.

Fifth, even if there were a cause of the universe, it does not follow that the cause is loving, moral, conscious, prayer-answering, or connected to human salvation.

The religious excitement over the Big Bang comes from adding theology to physics.

The science itself is more modest.

It says the observable universe has expanded from an early hot, dense condition. It does not say God commanded it into existence.

This is one of the repeated mistakes in religious argument.

Religion takes a scientific discovery and claims ownership of its mystery.

But mystery is not evidence for God.

It is a frontier for further investigation.

What the Big Bang Explains Well

The Big Bang model explains several major features of the universe.

First, it explains cosmic expansion.

If the universe is expanding now, then it was denser in the past. The Big Bang model gives a framework for understanding that expansion and tracing cosmic history backward.

Second, it explains the cosmic microwave background radiation.

The early universe was once so hot and dense that light could not travel freely. Later, as the universe expanded and cooled, atoms formed, and light was able to move through space. That ancient light still fills the universe today as a faint microwave glow.

This is one of the strongest pieces of evidence in modern cosmology.

Third, the Big Bang model helps explain the abundance of light elements, especially hydrogen and helium. The early universe provided conditions in which these light elements could form in predictable amounts.

Fourth, it helps explain the growth of cosmic structure. Tiny early variations in density eventually grew, under gravity, into galaxies and clusters of galaxies.

These are not vague claims.

They are measurable.

They are testable.

They are part of a scientific framework that can be refined, corrected, and improved.

This is why Big Bang cosmology is powerful.

It does not simply say “the universe began.”

It explains observed features of the universe.

That is what real explanations do.

They connect evidence.

They make predictions.

They invite testing.

They improve with measurement.

They can be challenged.

They can be corrected.

A sacred creation story does none of this.

It may comfort. It may inspire. It may give identity to a community. But it does not measure cosmic background radiation or explain the abundance of helium.

Cosmology does.

Light From the Infant Universe

One of the most beautiful ideas in modern science is that the young universe left light behind.

Not a divine message.

Not a sacred sentence.

Not a word spoken by God.

Light.

Radiation.

A physical trace.

In the early universe, matter and radiation were packed into a hot, dense state. Light could not travel freely because it was constantly scattered. As the universe expanded, it cooled. Eventually, atoms formed, and light could move through space. Over billions of years, that light stretched with the expansion of the universe and became the cosmic microwave background.

This is sometimes called the afterglow of the Big Bang.

The phrase is poetic, but the evidence is physical.

This ancient radiation fills the universe. It can be measured. It has a temperature. It contains tiny variations that help explain how later cosmic structures developed.

This is a powerful point.

The early universe did not leave behind a religious signature.

It left behind data.

Science learned to read that data.

That is why evidence-based cosmology is stronger than creation by faith. It does not ask us to believe an ancient story because it is sacred. It shows us measurable traces of cosmic history.

A creation story says, "Believe."

Cosmology says, "Look."

That difference matters.

What the Big Bang Does Not Explain Fully

A strong book must be honest.

The Big Bang model does not explain everything.

It does not give a complete answer to why there is something rather than nothing.

It does not fully explain the deepest origin of physical law.

It does not yet unite general relativity and quantum mechanics into a complete theory of quantum gravity.

It does not tell us with certainty what happened at the earliest boundary where our current theories become incomplete.

It does not fully explain dark matter.

It does not fully explain dark energy.

It does not solve every question about cosmic inflation.

It does not settle every debate about time, causation, or the ultimate foundation of reality.

Science should admit these limits.

But these limits do not help religion in the way many believers assume.

A gap in scientific knowledge is not evidence for God.

If science has not yet answered a question, the honest conclusion is not "therefore God." The honest conclusion is "we do not yet know."

This is a more disciplined position than faith.

Faith often fills the unknown with certainty.

Science leaves the unknown open until evidence improves.

The fact that cosmology has unanswered questions does not make ancient creation stories true. It does not prove a creator. It does not identify a divine mind. It does not justify worship, scripture, prayer, miracles, heaven, or hell.

Unanswered questions remain unanswered.

They do not become evidence for God simply because they are difficult.

Beginning Is a Dangerous Word

The word beginning creates confusion.

In ordinary life, a beginning happens in time.

A movie begins at 8 PM.

A child's school day begins in the morning.

A storm begins when clouds gather and rain starts.

A fire begins when fuel ignites.

In each case, there is a before and an after.

But the universe is different.

If time itself is part of the universe, then asking what happened before the universe may not be like asking what happened before a movie or a storm.

This is hard to understand because human language evolved for ordinary life, not cosmic origin.

When people hear "the beginning of the universe," they imagine a first moment sitting inside a larger time. They imagine time already flowing, then the universe appearing within it. But if time is part of the universe, the idea of "before the universe" becomes more complicated.

This does not mean there are no questions to ask.

It means our usual way of asking may be inadequate.

Religious arguments often depend on ordinary language:

The universe began.

Everything that begins has a cause.

Therefore the universe has a cause.

That cause is God.

But if time and causation are part of the physical universe, then the argument is not as simple as it sounds.

The beginning of the universe may not be like the beginning of a house, a painting, or a machine.

The universe may not require a maker in the way a chair requires a carpenter.

A beginning in cosmology is a scientific and philosophical challenge.

It is not automatic evidence for God.

The Big Bang and the Word "Nothing"

Another common confusion involves the word nothing.

People often say, "The universe came from nothing."

Then others respond, "Nothing comes from nothing, so God must have created it."

But the word nothing has different meanings.

In ordinary speech, nothing may mean empty space. But empty space is not absolute nothing. It can have physical properties. It can contain fields, energy, laws, and quantum activity.

In philosophy, absolute nothing would mean no space, no time, no matter, no energy, no laws, no fields, no possibilities, no anything.

These are very different ideas.

When physicists discuss the early universe, quantum vacuum, or cosmological models, they are usually not speaking about absolute philosophical nothing in the simple everyday sense. They are speaking within mathematical and physical frameworks.

This distinction matters.

If someone says physics explains how a universe could arise from "nothing," we must ask what kind of nothing is meant.

If someone says God created the universe from nothing, we must also ask what that means.

How does a timeless, spaceless, immaterial God act?

How does divine will produce physical reality?

What mechanism connects God's intention to matter, energy, space, and time?

If God existed, then there was not absolute nothing. There was God.

So the religious answer does not truly begin with nothing either.

It begins with a very complex something: God.

This is why “nothing” must be handled carefully.

Careless language creates false arguments.

Does the Big Bang Prove Genesis?

Some believers argue that the Big Bang proves the biblical creation story.

This claim is not convincing.

Genesis was written in an ancient religious and cultural context. It is not a physics textbook. It does not describe cosmic microwave background radiation, redshift, nucleosynthesis, galaxy formation, dark matter, dark energy, or space-time expansion. It does not provide mathematical models. It does not make testable predictions about cosmology.

It speaks in the language of ancient mythic imagination.

The same is true of other religious creation stories. They may be meaningful to believers. They may have literary, cultural, and historical importance. But they are not modern cosmology.

If the Big Bang supports one scripture, which scripture does it support?

The Bible?

The Quran?

The Vedas?

Other creation myths?

Many religious traditions claim creation. They cannot all be uniquely confirmed by the same scientific evidence.

This shows the problem.

Religions often reinterpret scientific discoveries after the fact and claim that their scriptures knew it all along. But if a text needs modern science to be reinterpreted as science, then the science did the work.

Scripture did not discover cosmic expansion.

Astronomy did.

Scripture did not predict the cosmic microwave background.

Physics did.

Scripture did not measure the age of the universe.

Cosmology did.

The Big Bang does not prove Genesis.

It shows that ancient scripture was not the source of modern cosmic knowledge.

The Priest Who Helped Cosmology

Georges Lemaître is important in this story.

He was a Belgian physicist and a Catholic priest. He proposed an expanding universe model and developed the idea of a primeval atom, an early form of what later became Big Bang cosmology.

His life is sometimes used by believers to show that science and faith can coexist.

That is true in one sense.

A religious person can do good science. A priest can make a scientific contribution. A believer can understand physics. The history of science is full of religious people who advanced natural explanations.

But this fact does not prove God.

Lemaître's scientific model did not depend on scripture. It did not prove Catholic doctrine. It did not show that prayer works. It did not identify a personal creator.

His physics stood on mathematical and observational grounds.

Indeed, Lemaître was careful about separating science and theology. He did not want his cosmology turned too quickly into religious proof.

This is a valuable lesson.

The private belief of a scientist is not the same as the explanatory content of the science.

Newton believed in God, but gravity works without God.

Boyle was religious, but chemistry explains matter without magic.

Pasteur had religious belief, but germ theory explains infection without demons.

Lemaître was a priest, but expanding universe cosmology explains cosmic history without divine command.

Science can be done by believers.

But the scientific explanation still does not need God.

A Scientific Model Is Not Sacred Doctrine

Another important point must be understood.

The Big Bang model is not sacred doctrine.

It can be revised.

It has already developed through many stages. New observations have refined it. Inflationary models were introduced to address certain problems. Measurements of the cosmic microwave background improved the details. Dark matter and dark energy became part of the standard cosmological model. Debates continue about the

Hubble constant, early universe physics, and the deeper nature of cosmic origin.

This flexibility is not weakness.

It is strength.

Science changes when evidence demands change.

Religion often treats change as danger. Science treats correction as progress.

If the Big Bang model is improved, expanded, or modified, science will not collapse. It will grow. The goal is not to defend one sacred sentence. The goal is to understand reality as well as evidence allows.

That is very different from religious certainty.

A sacred text must be protected.

A scientific model must be tested.

A doctrine must be defended.

A theory must be corrected.

This is why science is a better path to truth.

The Big Bang is powerful not because it is final, but because it is evidence-based and open to improvement.

God explanations usually do not improve in that way.

They retreat, reinterpret, and survive emotionally.

Science explains more.

Why God Adds Nothing to Big Bang Cosmology

Suppose someone says, "God caused the Big Bang."

What does this add to the science?

Does it change the equations?

Does it improve the measurement of cosmic expansion?

Does it explain the cosmic microwave background?

Does it predict the abundance of light elements?

Does it clarify dark matter?

Does it solve dark energy?

Does it explain inflation?

Does it tell us why physical laws have the form they do?

Does it identify what happened at the earliest boundary of our models?

Does it tell us how a timeless being creates time?

Does it provide a testable prediction?

No.

The sentence “God caused the Big Bang” may satisfy a believer emotionally, but scientifically it adds nothing.

It is not a model.

It is not an explanation.

It is a theological label placed over a scientific question.

This is the same pattern we have seen again and again.

God explained thunder until electricity explained thunder.

God explained disease until microbes explained disease.

God explained species until evolution explained species.

God explained planetary motion until gravity explained planetary motion.

Now God is placed at the Big Bang.

But the method is the same.

Find the gap.

Insert God.

Call it explanation.

Science-based atheism refuses this move.

It does not say we know everything.

It says we should not call ignorance God.

The Strongest Religious Reply

A thoughtful believer may say:

“The Big Bang strongly supports God. If the universe had a beginning, it cannot explain itself. Something beyond space, time, matter, and energy must have caused it. That cause must be timeless, spaceless, immaterial, powerful, and creative. This sounds like God.”

This is probably the strongest religious use of the Big Bang.

It deserves serious attention.

But it still fails as proof.

First, the Big Bang model does not necessarily describe creation from absolute nothing. It describes the early hot, dense state of the observable universe and its expansion.

Second, even if the universe had a beginning, the cause of that beginning is not automatically personal or divine.

Third, describing the cause as timeless, spaceless, immaterial, and powerful does not give us the God of religion. It does not give love, morality, prayer, heaven, scripture, miracles, or salvation.

Fourth, saying God is the cause raises the question of why God does not need a cause.

Fifth, if God can be uncaused, perhaps the deepest reality behind the universe can also be uncaused.

Finally, the religious argument depends on concepts like cause and beginning that may not apply simply to the earliest universe.

Therefore, the Big Bang does not prove God.

It raises deep questions.

But deep questions are not divine answers.

Evidence-Based Response

The evidence-based response is clear.

The Big Bang is one of the greatest scientific models in human history. It explains the expansion of the universe, the cosmic microwave background, the abundance of light elements, and the broad development of cosmic structure.

But it does not prove God.

It does not prove creation from nothing.

It does not prove a personal creator.

It does not prove scripture.

It does not prove divine intention.

It does not prove that the universe was made for human beings.

It shows that the universe has a physical history.

That history is better studied through evidence, mathematics, observation, and scientific humility than through ancient creation stories.

Science-based atheism does not need to deny mystery.

It only refuses to turn mystery into God.

That refusal is not arrogance.

It is honesty.

Chapter Summary

The Big Bang is often misunderstood as a religious creation event. Many people imagine it as God lighting a cosmic fuse or as an explosion into empty space. This is not what the science says.

Big Bang cosmology describes the expansion and cooling of the observable universe from an earlier hot, dense state. It is supported by evidence such as cosmic expansion, cosmic microwave background radiation, light element abundance, and large-scale structure.

The Big Bang does not answer every question. It does not fully explain why there is something rather than nothing. It does not yet provide a complete theory of the earliest possible state of reality. But these limits do not prove God.

A mystery is not evidence.

A beginning is not a creator.

The Big Bang is cosmology, not theology.

Key Points

The Big Bang was not an explosion into empty space.

It describes the expansion of space itself from an early hot, dense state.

Big Bang cosmology is based on evidence, not scripture.

It explains cosmic expansion, background radiation, light element abundance, and large-scale structure.

The cosmic microwave background is physical evidence, not a divine message.

The Big Bang does not prove creation from absolute nothing.

The Big Bang does not prove Genesis or any other scripture.

A beginning, even if accepted, is not automatically God.

God adds no mechanism, prediction, or testable explanation to Big Bang cosmology.

Science-based atheism accepts cosmic mystery without turning it into divine certainty.

Transition to the Next Chapter

The Big Bang is not a religious creation story.

It is a scientific model.

But why should we trust it?

The answer is evidence.

Modern cosmology does not rest on imagination alone. It rests on observation, prediction, measurement, and repeated confirmation. One of its greatest pieces of evidence is the faint afterglow of the early universe: the cosmic microwave background.

In the next chapter, we will examine how science learned to read that ancient light, and why evidence-based cosmology is far stronger than creation by faith.

Sources

Georges Lemaître, writings on the expanding universe and primeval atom

Edwin Hubble, "A Relation Between Distance and Radial Velocity Among Extra-Galactic Nebulae"

Steven Weinberg, *The First Three Minutes*

Stephen Hawking, *A Brief History of Time*

Alan Guth, *The Inflationary Universe*

Sean Carroll, *From Eternity to Here*

NASA resources on Big Bang cosmology and cosmic expansion

Stanford Encyclopedia of Philosophy, “Cosmology and Theology”

Chapter 6: The Ancient Light That Confirmed a Hot Early Universe

The Ancient Light Still Around Us

Long before Earth existed, before oceans, trees, animals, language, temples, scriptures, or gods, the young universe was glowing.

That glow is still here.

It is not visible to our eyes. We do not see it when we look at the night sky. We do not feel its warmth on our skin. It does not speak. It does not bless. It does not command. It does not tell us that a creator made the universe.

But it fills space.

It surrounds us from every direction.

It is a physical trace from the early universe, stretched and cooled by cosmic expansion over billions of years.

This ancient light is called the cosmic microwave background.

It is one of the strongest pieces of evidence for Big Bang cosmology.

To understand its importance, we must understand what science is doing here. Science is not asking people to believe a sacred story. It is not saying, "Trust this because an ancient book says so." It is not saying, "Accept this because a priest, prophet, or tradition declares it."

Science says something more powerful:

Look at the evidence.

Measure it.

Test the prediction.

Compare the model with reality.

Correct the model when needed.

This is why modern cosmology is so different from religious creation stories. A creation story may be emotionally moving. It may shape culture. It may give identity and comfort. But it cannot measure the cooled radiation from the early universe.

Cosmology can.

The Big Bang model became strong not because it sounded dramatic, but because the universe itself carried traces that matched the model.

The ancient light was still there.

Science learned to read it.

From Sacred Story to Measured Evidence

Human beings have always told stories about beginnings.

In one story, God speaks and the world appears.

In another, the universe emerges from cosmic waters.

In another, creation comes from a divine egg, a cosmic sacrifice, a primal being, or a battle among gods.

These stories are humanly interesting. They tell us much about the cultures that created them. They reveal fear, hope, geography, power, imagination, and the human need for order.

But they do not provide scientific evidence.

They do not predict background radiation.

They do not calculate elemental abundance.

They do not explain galactic redshift.

They do not describe the growth of structure from early density fluctuations.

They do not offer measurements that can be checked by instruments.

Big Bang cosmology is different.

It is not a story preserved because it is sacred.

It is a model supported because evidence points toward it.

This difference is critical.

Religion often begins with authority. A scripture says. A prophet says. A tradition says. A priest says. A community repeats. A child inherits.

Science begins with reality. What is observed? What is measured? What follows from the model? What evidence would confirm it? What evidence would weaken it?

This is why science-based atheism is stronger than religious creation belief. It does not depend on inherited authority. It depends on reality itself.

The universe is not believed into existence.

It is investigated.

The Early Universe Was Opaque

To understand the cosmic microwave background, we need a simple picture of the early universe.

The young universe was extremely hot and dense. It was filled with particles, radiation, and energy. In that early state, light could not travel freely for long distances. Photons were constantly scattered by charged particles. The universe was like a glowing fog.

Then the universe expanded.

As it expanded, it cooled.

Eventually, conditions changed enough for electrons and nuclei to combine and form neutral atoms, mainly hydrogen and helium. Once that happened, light could travel much more freely. The universe became transparent.

That ancient light has been traveling through space ever since.

But as the universe expanded, the wavelength of that light stretched. Over billions of years, it cooled into microwave radiation.

That radiation is the cosmic microwave background.

This is a deeply natural explanation.

No angel carries the light.

No divine hand paints the pattern.

No scripture predicted the temperature map.

The radiation exists because the early universe had a physical history. It was hot. It expanded. It cooled. It changed from opaque to transparent. Light began to travel freely. Expansion stretched that light into microwaves.

This is not myth.

It is a measurable relic of the early universe.

The Prediction That Changed Cosmology

One reason the cosmic microwave background is so important is that it was predicted before it was clearly observed.

That matters greatly.

A scientific theory becomes stronger when it predicts something that is later found. Prediction is one of the marks of a serious explanation.

If the universe began in a hot, dense state, then some radiation from that early state should still remain. It should fill the universe in all directions. It should have cooled over time because of expansion. It should now be detected as a faint microwave background.

That is what scientists eventually found.

In 1965, Arno Penzias and Robert Wilson detected a persistent microwave signal that seemed to come from all directions. At first, they did not fully understand what they had found. The noise was stubborn. It would not go away. It was not a loose wire, a bad instrument, or an ordinary local disturbance.

The strange signal turned out to be a message from the early universe.

Not a divine message.

A physical one.

It matched what Big Bang cosmology expected: a faint afterglow of an early hot, dense universe.

This discovery was one of the great moments in modern science.

It did not prove God.

It did not confirm a religious creation story.

It confirmed that the universe had once been much hotter and denser than it is today.

That is the difference between religious imagination and scientific success.

Religion may say, “In the beginning, God created.”

Science says, “If the universe was once hot and dense, then we should observe a cooled background radiation. Let us look.”

When science looked, the evidence was there.

That is why the cosmic microwave background is so powerful.

It is not belief.

It is prediction meeting observation.

A Universe With a Temperature

The cosmic microwave background has a temperature.

That sentence itself is extraordinary.

The universe has a measurable thermal remnant from its early history. Today, this background radiation is extremely cold, only a few degrees above absolute zero. But it is not zero. It fills space as a faint glow from the early universe.

This should make us pause.

Creation stories speak in human language: command, will, purpose, heaven, earth, light, darkness, divine intention.

Cosmology speaks in measurable terms: temperature, wavelength, expansion, density, radiation, fluctuation, redshift.

This does not make cosmology less beautiful.

It makes it more reliable.

A temperature can be measured.

A claim can be tested.

An instrument can be built.

A result can be checked.

A map can be made.

This is how cosmology became a science.

The cosmic microwave background is not just a vague glow. It has been studied in great detail. It is remarkably uniform, but not perfectly uniform. Tiny variations in its temperature are crucial. Those slight differences are the seeds of later cosmic structure.

From small early variations, gravity helped matter gather. Over immense time, those regions became galaxies, clusters, and the large-scale structure of the universe.

In other words, the universe did not only leave an afterglow.

It left a pattern.

Science read the pattern.

The Seeds of Galaxies

The universe today is not smooth.

It contains galaxies, clusters of galaxies, filaments, walls, and great cosmic voids. Matter is arranged in large-scale structure. This structure did not appear by magic. It developed from earlier conditions.

The cosmic microwave background contains tiny variations in temperature and density. These small differences mattered enormously. Over time, gravity amplified them. Regions with slightly more matter attracted more matter. Slowly, structure grew.

Galaxies did not need to be placed by a divine hand.

They developed through natural processes.

Gravity shaped the cosmic web.

Gas collapsed.

Stars formed.

Galaxies grew and merged.

Clusters assembled.

The universe became structured over time.

This is another powerful example of natural explanation.

A religious mind may see a galaxy and say, "God made it."

Science asks, "How did it form?"

That question leads to better answers.

It leads to gravity, dark matter, gas dynamics, star formation, mergers, and cosmic history.

The scientific answer may be less simple than "God made it," but it is far more informative.

It explains process.

And process is what religion so often lacks.

COBE, WMAP, and Planck

The study of the cosmic microwave background did not stop with discovery.

Later missions measured it with increasing precision.

COBE, the Cosmic Background Explorer, provided important measurements showing that the cosmic microwave background has the expected blackbody spectrum and tiny temperature variations.

WMAP, the Wilkinson Microwave Anisotropy Probe, mapped those tiny variations in greater detail.

Planck, a European Space Agency mission, measured the cosmic microwave background with even higher precision, giving scientists a more detailed picture of the early universe.

These missions changed cosmology.

They allowed scientists to estimate the age, composition, geometry, and development of the universe with remarkable accuracy. They helped support what is often called the standard model of cosmology.

This is not mythology.

This is measurement.

A creation story remains fixed because a community preserves it.

A scientific model improves because instruments improve.

That difference matters.

If a telescope becomes more powerful, the model must face better evidence.

If a satellite measures more accurately, the theory must survive more precise testing.

If new data disagree with the model, the model must be revised.

That is why science is trustworthy. Not because scientists are perfect. They are not. But because the method allows reality to correct human beings.

Religion often asks reality to fit the story.

Science asks the story to fit reality.

The cosmic microwave background is one of the greatest examples of that discipline.

Light Elements and the Early Universe

The cosmic microwave background is not the only evidence for Big Bang cosmology.

Another important line of evidence comes from the abundance of light elements.

In the early universe, conditions were hot and dense enough for certain nuclear reactions to occur. This process, called Big Bang nucleosynthesis, produced mainly hydrogen and helium, with small amounts of deuterium, helium-3, and lithium.

The predicted amounts broadly match what scientists observe.

This is important because it gives an independent line of evidence. The Big Bang model does not depend on only one observation. It connects many observations into one framework.

Expansion.

Background radiation.

Light element abundance.

Large-scale structure.

Cosmic age.

The development of galaxies.

A good scientific model does this. It brings separate facts together.

Religion can give one sweeping answer: God created all things. But that answer does not explain the details. It does not tell us why helium has the abundance we observe. It does not explain why the cosmic background has a particular temperature. It does not predict density fluctuations.

Science does.

This is why saying "God did it" is weak. It explains everything in a general way, and therefore explains nothing in a specific way.

Big Bang cosmology is not complete, but it is specific.

It gives details.

Details are where real explanation lives.

The Standard Cosmological Model

Modern cosmology is more than the phrase Big Bang.

It includes a broader scientific framework often called the standard model of cosmology. In simple terms, this model brings together cosmic expansion, the cosmic microwave background, light element abundance, large-scale structure, dark matter, dark energy, and the evolution of galaxies.

This framework is not perfect.

It is not final.

But it is powerful because many different kinds of evidence point toward the same broad picture.

The universe expanded from an earlier hot, dense state.

It cooled.

Matter formed.

Light began to travel freely.

Small variations grew into large structures.

Stars and galaxies formed.

The expansion of the universe changed over time.

Dark matter appears to have helped shape cosmic structure.

Dark energy appears connected to the accelerated expansion of the universe.

This is an extraordinary scientific achievement.

No ancient scripture produced this picture.

No prayer revealed it.

No prophet measured it.

It came from telescopes, satellites, mathematics, physics, observation, correction, and generations of human inquiry.

This is exactly why science-based atheism is not merely disbelief in God. It is trust in the method that actually discovers reality.

Religion gave humanity stories.

Science gave humanity the cosmos.

Cosmic Inflation: A Deeper Early Chapter

One major idea in modern cosmology is inflation.

Inflation is the proposal that the very early universe underwent an extremely rapid expansion. It was introduced to help explain several puzzles, including why the universe appears so uniform on large scales and why its geometry appears so close to flat.

This is a good example of how science works.

Scientists did not say, "The universe looks uniform, therefore God made it that way."

They asked what physical process might explain the evidence.

Inflation may be part of the answer.

It may also need revision. Some versions are debated. The details remain under study. Scientists continue to test models against evidence from the cosmic microwave background and other observations.

This uncertainty should not embarrass science.

It should impress us.

Science does not protect ideas because they are sacred. It tests them. If inflation is correct, it will become stronger through evidence. If it is incomplete, it will be revised. If it is wrong, better models will replace it.

That is intellectual honesty.

God explanations do not usually improve in this way. They remain vague, immune, and flexible.

Inflation shows that when science meets a problem, it builds models and seeks evidence.

It does not kneel before mystery.

Dark Matter, Dark Energy, and Honest Mystery

A serious book on cosmology must not pretend that everything is fully understood.

It is not.

Modern cosmology includes major mysteries.

Dark matter appears to affect the motion of galaxies and the formation of large-scale structure, but its exact nature remains unknown.

Dark energy appears connected to the accelerated expansion of the universe, but its deeper nature remains uncertain.

The Hubble tension, the difference between some measurements of the expansion rate of the universe, remains an active scientific problem.

These are real questions.

Science should admit them.

But admitting mystery is not surrendering to God.

This is one of the most important lessons in this book.

The unknown is not owned by religion.

When science says, “We do not yet know what dark matter is,” that does not mean “therefore God.” It means we do not yet know what dark matter is.

When science says, “We do not fully understand dark energy,” that does not mean “therefore God.” It means more research is needed.

When scientists debate the Hubble tension, that does not mean cosmology has failed. It means measurement has become precise enough to reveal new problems.

This is how science advances.

Mystery is not a defeat for science.

Mystery is the beginning of science.

Religion often treats mystery as a home for God.

Science treats mystery as work to be done.

The Strength of Multiple Lines of Evidence

One reason Big Bang cosmology is powerful is that it rests on multiple lines of evidence.

Cosmic expansion points to a universe that was denser in the past.

The cosmic microwave background points to a hot early universe.

Light element abundance supports early nuclear processes.

Large-scale structure shows how small early variations grew over time.

Observations of distant galaxies reveal a universe that changes with age.

These independent lines of evidence support one another.

That is how science becomes strong.

A single observation may be misunderstood. A single measurement may be wrong. A single argument may have weaknesses. But when many different kinds of evidence point in the same direction, confidence grows.

This is very different from religious creation belief.

Religious belief often begins with a conclusion and interprets everything to fit it.

Science builds from evidence upward.

This is why science-based atheism is not merely disbelief. It is a disciplined way of understanding reality. It does not reject God because it dislikes religion. It rejects God because natural explanations repeatedly do the work better.

Big Bang cosmology does not need God because it has evidence.

Expansion.

Radiation.

Elements.

Structure.

Mathematics.

Measurement.

Correction.

These are stronger than revelation.

Why Evidence Matters More Than Feeling

Many people feel God when they look at the universe.

They see stars and feel design.

They see order and feel mind.

They see beauty and feel purpose.

They see the cosmic background of existence and feel that something greater must be there.

The feeling is real.

But feeling is not evidence.

This distinction is essential.

A person may feel that the Sun moves around Earth. It does not.

A person may feel that Earth is still. It moves.

A person may feel that a dead loved one is present in the room. The feeling may be emotionally real, but it does not prove an external spirit.

A person may feel that the universe must have been created by God. That feeling may be powerful, but it does not prove God.

Science protects us from the errors of feeling.

It asks for evidence that can be examined by others. It asks for measurement. It asks for predictions that can fail. It asks for correction.

This does not make science cold.

It makes science honest.

In cosmology, honesty matters because the questions are enormous. If we allow feeling to become proof, then every religious tradition can claim the universe confirms its own God. That leads nowhere.

Evidence gives us a shared path.

The cosmic microwave background does not depend on Hindu, Christian, Muslim, Jewish, Buddhist, or atheist feeling. It can be measured by anyone with the right tools and methods.

That is why science is universal in a way religion is not.

Does Evidence-Based Cosmology Leave Room for God?

Some believers may say that evidence-based cosmology does not exclude God. Perhaps God created the early universe. Perhaps God designed the laws. Perhaps God set the constants. Perhaps God works through the Big Bang rather than against it.

This position is possible.

But it is not required.

That distinction matters.

The fact that a belief can be made compatible with science does not mean science supports the belief.

A person can say God works through cosmic expansion. But cosmic expansion does not show God.

A person can say God created the cosmic microwave background. But the radiation itself does not identify divine action.

A person can say God designed the laws behind Big Bang cosmology. But that claim gives no additional measurement, no better model, no clearer prediction.

It is belief added to science, not science itself.

This is the retreat of God in modern form.

God is not used to replace the scientific explanation. God is placed behind it. But once God is placed behind the explanation, God no longer explains the phenomenon directly.

The evidence does not require God.

And if the evidence does not require God, science-based atheism has the stronger position.

It does not add unnecessary beings.

It does not multiply mysteries.

It accepts the natural explanation as far as the evidence supports it.

The Religious Reply

A thoughtful believer may say:

"The evidence for the Big Bang supports the idea that the universe had a beginning. The cosmic microwave background, expansion, and light element abundance all point to a universe with a finite past. If the universe had a beginning, then it still needs a creator. Science may describe how the universe developed after the beginning, but God explains why there was a beginning at all."

This is a serious reply.

It correctly recognizes that cosmology raises deep questions. It also correctly notes that science has not fully answered every question about ultimate origin.

But the reply still goes beyond the evidence.

The cosmic microwave background supports an early hot, dense state. It does not support a divine command. The expansion of the universe supports physical cosmology. It does not identify a creator. The abundance of light elements supports early nuclear processes. It does not reveal a mind behind those processes.

Saying "God explains why" may feel satisfying, but it does not actually explain how a timeless, spaceless, immaterial mind creates a physical universe.

It gives no testable model.

It gives no prediction.

It gives no measurement.

It gives no way to distinguish one God from another.

It gives no evidence that the creator is loving, personal, moral, or prayer-answering.

It simply places God at the edge of current knowledge.

That is not explanation.

It is theological placement.

Evidence-Based Response

The evidence-based response is clear.

Big Bang cosmology is powerful because it is supported by multiple lines of evidence. Cosmic expansion, the cosmic microwave background, light element abundance, and large-scale structure all point toward a universe that was hotter, denser, and very different in the past.

This is one of the great achievements of science.

But none of this proves God.

The evidence supports a physical history of the universe. It does not support a supernatural creator. It does not identify divine intention. It does not prove scripture. It does not show that the universe was made for human beings.

God can be added as belief, but God is not required by the evidence.

That is the central point.

Science-based atheism does not deny that mysteries remain. It insists that mysteries should not be turned into gods.

The cosmic microwave background is not the echo of a divine voice.

It is the afterglow of a hot early universe.

That explanation is enough.

Chapter Summary

Big Bang cosmology is not speculation or religious myth. It is supported by strong evidence.

The cosmic microwave background is one of the most important pieces of evidence. It is the cooled radiation left from the early universe, when the universe became transparent and light began traveling freely. Its discovery and later detailed measurement strongly supported the Big Bang model.

Other evidence also supports modern cosmology, including cosmic expansion, light element abundance, large-scale structure, inflationary models, and the standard cosmological framework.

Science remains honest about what it does not yet know. Dark matter, dark energy, inflation, quantum gravity, and the Hubble tension remain active areas of study. But unanswered questions are not evidence for God.

Evidence-based cosmology shows that the universe has a natural history.

God is not required to explain that history.

Key Points

Big Bang cosmology is supported by evidence, not faith.

The cosmic microwave background is ancient radiation from the early universe.

Its discovery strongly supported the hot Big Bang model.

COBE, WMAP, and Planck measured the cosmic microwave background in increasing detail.

Light element abundance supports early universe nuclear processes.

Large-scale structure grew from small early density variations.

Modern cosmology includes expansion, dark matter, dark energy, inflation, and precision measurements.

Dark matter and dark energy remain mysteries, but mystery is not evidence for God.

Multiple lines of evidence make cosmology strong.

God can be added as belief, but God is not required by the evidence.

Transition to the Next Chapter

The evidence for Big Bang cosmology is powerful.

The universe has expanded. It has cooled. It has changed. Its early light still fills space. Its ancient structure can be measured. Its history can be reconstructed.

But one question remains unavoidable.

If the universe has a history, does that history need a first cause?

This is one of the oldest and strongest arguments for God. It sounds simple, but it is not. It brings us to the difficult questions of time, causation, existence, and explanation.

In the next chapter, we will ask: does the universe need a first cause?

Sources

Arno Penzias and Robert Wilson, writings on the discovery of the cosmic microwave background

George Gamow, Ralph Alpher, and Robert Herman, writings on early Big Bang predictions

Steven Weinberg, *The First Three Minutes*

Alan Guth, *The Inflationary Universe*

Sean Carroll, *The Big Picture*

Brian Greene, *The Fabric of the Cosmos*

NASA resources on Big Bang cosmology, WMAP, dark matter, and dark energy

ESA resources on Planck and cosmic microwave background radiation

Chapter 7: Does the Universe Need a First Cause?

The Question That Sounds So Simple

Every child eventually asks the question that philosophy has never fully escaped:

What caused everything?

The question feels natural because everything around us seems to have a cause. A child exists because parents existed. A house exists because builders built it. A book exists because someone wrote it. A tree grows because a seed fell into soil. A fire begins because something ignited it. A wound exists because something injured the body.

So the mind moves naturally to the largest question:

If everything has a cause, what caused the universe?

For many people, the answer feels obvious.

God.

God caused the universe. God made space, time, matter, energy, stars, planets, and human life. God stands before all things, behind all things, beyond all things. The universe cannot explain itself, so God must explain the universe.

This argument is powerful because it gives the mind a stopping point. It seems to rescue us from the discomfort of endless questions. It gives mystery a name. It turns cosmic uncertainty into a familiar answer.

But a name is not necessarily an explanation.

Calling the mystery God does not automatically solve the mystery.

In fact, the first-cause argument may be one of the most difficult arguments in the entire debate about God. It touches the meaning of time, causation, existence, nothingness, physical law, and explanation itself.

It should not be mocked.

It should not be dismissed cheaply.

But it should also not be accepted too quickly.

The central question is not whether the universe raises deep mysteries.

It does.

The question is whether those mysteries prove God.

They do not.

Why the First-Cause Argument Feels Powerful

The first-cause argument feels powerful because it begins with ordinary experience.

In daily life, causes are everywhere.

A glass breaks because it falls.

A patient develops pneumonia because infection enters the lungs.

A car stops because the fuel is gone or the engine fails.

A river floods because rain is heavy and drainage is overwhelmed.

A baby cries because she is hungry, wet, frightened, or in pain.

Human beings learn from birth that things happen for reasons. The brain searches for cause because cause helps survival. If a noise comes from the bushes, we want to know what caused it. If a child has fever, we want to know why. If crops fail, we look for drought, pests, disease, soil, or weather.

Cause matters.

So when we look at the universe, we naturally ask for a cause.

This is not foolish. It is deeply human.

The problem begins when we take a rule learned inside the universe and apply it to the universe as a whole without caution.

Everything we experience happens within space and time. Causes in ordinary life are events inside the universe. One event happens, then another follows. A match is struck, then a flame appears. A stone hits glass, then the glass breaks. A virus infects cells, then illness may develop.

But the universe is not simply another object sitting inside a larger known room.

The universe is the total physical reality we are trying to explain.

It includes space, time, matter, energy, and the laws of physics as we know them.

So when we ask, “What caused the universe?” we are no longer asking an ordinary question. We are asking whether the entire framework in which ordinary cause operates has an explanation beyond itself.

That may be a meaningful question.

But it is not simple.

The Trap Hidden in a Simple Argument

The traditional first-cause argument is easy to state.

The universe began.

Everything that begins must have a cause.

The universe must have a cause.

That cause must be beyond the universe.

That cause is God.

The argument sounds clean. It has the rhythm of logic. It moves step by step. It feels almost unavoidable.

But the trap is hidden in the last step.

Even if the universe has a cause, why should that cause be the God of religion?

Why should it be personal?

Why should it be loving?

Why should it hear prayer?

Why should it care about human beings?

Why should it command worship?

Why should it reveal scriptures?

Why should it judge sin?

Why should it promise heaven or threaten hell?

A first cause, by itself, gives none of this.

At most, it gives a mysterious source, foundation, or explanation.

Religion then adds a great deal.

That addition is not proven by the argument.

This is one of the most important points in this chapter. Many religious arguments begin with a broad philosophical idea, then quietly attach to it the God people already worship. But the bridge between "first cause" and "personal God" is weak.

A cause is not automatically a creator.

A creator is not automatically a person.

A person is not automatically loving.

A loving creator is not automatically the God of any scripture.

The argument moves too quickly.

Cause Inside the Universe and Cause of the Universe

The difference between cause inside the universe and cause of the universe is crucial.

Inside the universe, causes are familiar. They happen in time. A cause comes before an effect, or at least stands in a physical relation to it. We can often observe, measure, test, and repeat causal relationships.

A spark causes fire.

Gravity causes a stone to fall.

Bacteria cause certain infections.

A blocked artery causes damage to heart muscle.

The cause and effect both occur within the universe.

But when we ask about the cause of the universe, we are asking something different. The universe is not one event among other events in a larger laboratory. It is the total physical reality within which our ordinary ideas of cause operate.

This creates a problem.

If time itself is part of the universe, then what does it mean to say something caused the universe before time existed?

The word before may not apply in the ordinary way.

If there was no time before the universe, then a cause before the universe becomes difficult to understand.

Some theologians respond by saying God caused the universe timelessly. But this creates new questions. What does timeless causation mean? How does a timeless being choose to create? Does choice require before and after? Does intention require sequence? How can an unchanging timeless being cause a changing universe?

These questions do not make theology impossible.

But they show that the religious answer is not simple.

The believer often says, “The universe needs a cause.”

But when asked how God causes without time, the answer becomes mysterious.

So the mystery has not disappeared.

It has moved.

If Time Began, What Does Before Mean?

The idea that time may be connected to the universe is one of the most important challenges to simple creator arguments.

In ordinary life, every beginning happens within time.

A song begins after silence.

A race begins after the starting signal.

A fever begins after infection.

A day begins after night.

But the beginning of the universe, if that phrase is appropriate, is not like the beginning of a song or a race. It may involve the origin of time itself, or at least the boundary of time as described by our current physical theories.

So the question "What happened before the universe?" may be like asking, "What is north of the North Pole?"

The question uses familiar words, but the words may no longer fit.

This does not mean all questions disappear. It does not mean the universe explains itself completely. It does not mean science has the final answer to ultimate reality.

But it does weaken the simple version of the first-cause argument.

If time is part of the universe, then cause cannot be used in the ordinary time-before-time sense without careful explanation.

Religion often depends on ordinary mental pictures:

God existed first.

Then God created.

Then the universe began.

But if time itself is part of the universe, the word first becomes difficult.

There may be deeper ways to discuss causation, but they are no longer simple common sense. They are advanced philosophy and physics. And they do not automatically lead to the personal God of religion.

Aquinas in Plain Language

Thomas Aquinas gave several famous arguments for God.

One of his arguments begins with motion or change. Things change. Whatever changes is changed by something else. This cannot go back forever in an infinite chain, Aquinas argued. Therefore, there must be a first unmoved mover.

Another argument focuses on causation. Causes exist. Nothing causes itself. There cannot be an infinite chain of causes. Therefore, there must be a first cause.

Aquinas identified this first cause with God.

These arguments were intellectually powerful in their time. They still influence theology today.

But several points must be made.

First, Aquinas worked within a medieval philosophical framework very different from modern physics. His ideas about motion, causation, and existence were shaped by Aristotle, not by relativity, quantum physics, or modern cosmology.

Second, even if one accepts the need for a first cause or unmoved mover, the argument does not automatically produce the God of Christianity, Islam, Hinduism, Judaism, or any specific religion.

Third, the first cause does not necessarily have to be a person. It could be a deeper physical reality, a necessary structure, a law-like foundation, or something beyond our current concepts.

Fourth, saying "this is what we call God" does not prove that the first cause is the God people worship.

That is the leap.

A first cause is a philosophical idea.

The God of religion is a much richer claim.

The first does not automatically prove the second.

The Kalam Argument

The Kalam cosmological argument has become popular in modern religious apologetics.

Its basic form is:

Whatever begins to exist has a cause.

The universe began to exist.

Therefore, the universe has a cause.

Then the argument often continues: this cause must be timeless, spaceless, immaterial, powerful, and personal. Therefore, God exists.

The argument is attractive because it is simple and direct.

But each step needs careful examination.

First, "whatever begins to exist has a cause" is based on ordinary experience inside the universe. We do not know whether this principle applies to the universe as a whole, especially if time itself is part of the system.

Second, “the universe began to exist” depends on how we interpret cosmology. Big Bang cosmology points to an early hot, dense state of the observable universe. It does not necessarily prove creation from absolute nothing.

Third, “therefore the universe has a cause” may be reasonable as a philosophical suggestion, but the nature of such a cause remains unclear.

Fourth, calling the cause personal is a major additional step. Why personal? Why conscious? Why intentional? Why moral? Why not an impersonal physical reality? Why not a deeper law? Why not a quantum cosmological process? Why not a brute fact?

Fifth, even if the cause were personal, it still would not identify which religion is true.

The Kalam argument may raise a serious question.

It does not prove God.

It especially does not prove the God of prayer, scripture, miracle, heaven, hell, and worship.

Leibniz and the Question Why Anything Exists

Another important version of the argument comes from Gottfried Wilhelm Leibniz.

He asked a profound question:

Why is there something rather than nothing?

This may be the deepest question human beings ask.

It does not simply ask what came before the universe. It asks why existence exists at all. Why is there reality? Why are there laws? Why is there any world instead of no world?

Leibniz appealed to the principle of sufficient reason. For every fact, there must be an explanation for why it is so and not otherwise. The universe, he argued, cannot explain itself fully. Therefore, the explanation must lie in a necessary being, which he identified with God.

This argument is more subtle than a simple beginning argument.

It does not depend only on time.

It asks for ultimate explanation.

But again, difficulties remain.

Why must every fact have an explanation?

Could reality itself be the ultimate brute fact?

Could the deepest laws or structures be necessary?

Does calling God necessary actually explain God, or does it simply stop the questioning?

If God is necessary, how do we know that?

If God necessarily exists, why does this necessary being create a contingent universe?

Why this universe and not another?

And again, how does one move from necessary being to a personal God who cares about human beings?

Leibniz asked a powerful question.

But the answer "God" may not be as explanatory as it first appears.

Brute Facts and Stopping Points

Every worldview needs a stopping point.

A theist may stop with God.

An atheist may stop with reality, physical law, the universe, the multiverse, quantum fields, or some deeper natural foundation not yet fully understood.

The theist often says, "God explains why the universe exists."

But the atheist can ask, "What explains God?"

The theist may answer, "God is necessary."

But the atheist can ask, "How do you know?"

If we are allowed to say that God is necessary and uncaused, why are we not allowed to consider that the deepest natural reality may be necessary or uncaused?

This is not a trick question.

It exposes a serious asymmetry.

The religious argument often refuses to let the universe be unexplained, but then allows God to be unexplained. It demands an explanation for existence, then stops at a being whose existence is even more mysterious.

A timeless, spaceless, immaterial, all-powerful mind is not a simple explanation.

It is an enormous claim.

If we must accept one ultimate stopping point, science-based atheism prefers the stopping point closer to evidence: reality itself, not an invisible supernatural person.

This does not solve every mystery.

But it avoids multiplying mysteries unnecessarily.

God does not remove the final question.

God becomes the final question.

Does the First Cause Have to Be Personal?

Religious arguments often claim that the first cause must be personal.

Why?

The usual answer is that if the cause were impersonal and eternal, then the universe would also be eternal. To explain a universe that begins, the cause must choose to create. Choice requires a personal agent. Therefore, the first cause is personal.

This argument sounds clever, but it is not decisive.

First, it assumes that creation is like a decision made in time. But if time begins with the universe, there is no earlier moment when God decides.

Second, it assumes that impersonal causes cannot produce finite effects. This may be based more on ordinary intuition than deep physics.

Third, even if something like choice were involved, that does not automatically give us a loving or moral God.

Fourth, a personal timeless cause is difficult to understand. Persons usually have thoughts, intentions, awareness, memory, sequence, and change. A timeless person is not like any person we know.

This shows again that God is not a simple answer.

When religion says the first cause must be personal, it is often trying to bring the God of worship into a philosophical argument that does not actually establish such a being.

The universe may have an ultimate explanation.

But the personal God of religion is not established by first cause alone.

From First Cause to Religious God

Suppose we grant, for the sake of argument, that there is a first cause.

What follows?

Not much religiously.

A first cause does not tell us that Christianity is true.

It does not tell us that Islam is true.

It does not tell us that Hinduism is true.

It does not tell us that Judaism is true.

It does not tell us that any scripture is inspired.

It does not tell us that prayer works.

It does not tell us that miracles happen.

It does not tell us that the soul survives death.

It does not tell us that God is loving.

It does not tell us that God is just.

It does not tell us that heaven or hell exists.

It does not tell us that human beings are the purpose of creation.

This is a major weakness.

The first-cause argument often begins as philosophy and ends as religion. But the bridge between them is weak.

The believer may say, "The first cause is what I mean by God."

But that is not what most people mean by God.

Most people do not worship an abstract first cause.

They worship a personal being.

They pray to someone.

They ask for help.

They seek forgiveness.

They hope for life after death.

They believe God cares.

The first-cause argument does not give them that.

It gives, at most, a metaphysical mystery.

The Problem of a Timeless Creator

A timeless creator creates special problems.

If God is timeless, then God does not experience before and after. But creation seems to involve before and after. First no universe, then universe. First no creation, then creation.

How does that work without time?

If God is unchanging, then how does God change from not creating to creating?

If God eternally wills the universe, why is the universe not eternal?

If God's will is timeless, what makes the universe begin at one time rather than another?

If time itself begins with the universe, what does "at one time" even mean?

These are difficult questions.

Theology has offered various answers. Some say God's causation is timeless. Some say God creates time itself. Some say the universe depends on God at every moment rather than beginning after God. These may be interesting philosophical ideas.

But they are not scientific explanations.

They are not models that can be tested.

They do not produce measurable predictions.

They do not clarify the physical origin of the universe.

They are ways of protecting belief from contradiction.

Science-based atheism can say something simpler:

We do not yet know the ultimate nature of cosmic origin. But saying "timeless creator" does not make the mystery clearer.

It makes the mystery religious.

The Problem of an Immaterial Cause

A creator outside the universe is often described as immaterial.

But if God is immaterial, how does God produce matter?

In ordinary life, causes involve relations within the physical world. A hand pushes a cup. Heat boils water. Gravity moves planets. Chemicals react. Cells divide. Radiation transfers energy. Causes have mechanisms.

If God is outside matter, energy, space, and time, what is the mechanism of creation?

How does divine intention become physical law?

How does a nonphysical mind create particles, fields, space-time, energy, and cosmic expansion?

The believer may answer: God does it by divine power.

But that is not an explanation.

It is another phrase for the same mystery.

A scientific explanation gives structure. It tells us how something happens. It connects one fact to another. It makes predictions. It can be examined. "Divine power" does not do that.

This is why God is not a strong explanation in cosmology.

God is invoked precisely where mechanism is missing.

Science seeks the mechanism.

Religion replaces the mechanism with mystery.

The Strongest Religious Reply

A thoughtful believer may say:

"The universe cannot explain itself. Everything contingent needs an explanation. The chain of causes cannot go back forever. There must be a necessary foundation of all reality. This necessary foundation must be beyond space, time, matter, and energy. That is what we mean by God. Science may explain events within the universe, but it cannot explain why the universe exists at all."

This is the strongest form of the first-cause argument.

It deserves respect.

It rightly asks a deep question. It rightly recognizes that science may not yet have a complete answer to ultimate existence. It rightly challenges atheists not to pretend that cosmic origin is easy.

But it still does not prove God.

It does not show that the necessary foundation is personal.

It does not show that it is conscious.

It does not show that it is loving.

It does not show that it is morally perfect.

It does not show that it hears prayers.

It does not show that it revealed scriptures.

It does not show that it cares about human beings.

It does not show that it is one God rather than another.

It does not show that calling it God adds explanatory power.

The believer may say, "God is the necessary foundation."

But the atheist may reply, "That is a name attached to a mystery, not a demonstrated being."

The deeper truth may be natural, not divine.

Evidence-Based Response

The evidence-based response is careful but firm.

The first-cause argument raises real questions about existence, time, cause, and explanation. These questions should not be mocked. They are among the deepest questions human beings ask.

But the argument does not prove God.

Cause is a concept we understand from events within the universe. It may not apply simply to the universe as a whole. If time itself is part of the universe, the idea of a cause before the universe becomes difficult. Big Bang cosmology points to an early hot, dense state, but it does not prove creation from absolute nothing.

Even if one accepts some kind of first cause, that does not prove a personal God. It does not prove the God of religion. It does not prove prayer, scripture, miracles, morality, heaven, hell, or divine love.

The first-cause argument moves too quickly from mystery to God.

Science-based atheism does not claim to have solved every ultimate question. It only refuses to pretend that God is an answer when God explains nothing clearly.

The honest answer may be:

We do not yet know the deepest foundation of reality.

That is better than saying:

We do not know, therefore God.

Chapter Summary

The first-cause argument is one of the oldest and most powerful arguments for God. It begins with a simple idea: everything that begins has a cause, the universe began, and therefore the universe must have a cause.

But the argument is more difficult than it first appears.

Causes are known from events within the universe. Applying causation to the universe as a whole is not simple, especially if time itself is part of the universe. The Big Bang does not necessarily prove creation from absolute nothing. It points to an early hot, dense state of the observable universe.

Even if a first cause were accepted, it would not automatically be the God of religion. It would not prove a personal, loving, moral, prayer-answering God. It would not prove scripture, miracle, heaven, hell, or divine judgment.

The universe may have deep mysteries.

But mystery is not proof of God.

Key Points

The first-cause argument feels powerful because cause is familiar in ordinary life.

Cause inside the universe may not apply simply to the universe as a whole.

If time is part of the universe, the idea of "before the universe" becomes difficult.

Aquinas, Kalam, and Leibniz offer serious versions of the cosmological argument.

None of these arguments automatically proves a personal God.

Calling God necessary does not prove God exists.

A first cause, even if accepted, does not prove any specific religion.

God often becomes a name for mystery, not an explanation of mystery.

Science-based atheism accepts unanswered questions without turning them into God.

Transition to the Next Chapter

The first-cause argument leads to an unavoidable question.

If everything needs a cause, why does God not need one?

If the universe cannot simply exist, why can God simply exist?

If God can be eternal, necessary, or uncaused, why cannot the deepest natural reality be eternal, necessary, or uncaused?

This question is often asked by children, but it is not childish.

In the next chapter, we will examine one of the most important challenges to creator arguments:

Who created God?

Sources

Thomas Aquinas, *Summa Theologica*

Gottfried Wilhelm Leibniz, writings on the principle of sufficient reason

William Lane Craig, writings on the Kalam cosmological argument

David Hume, *Dialogues Concerning Natural Religion*

Bertrand Russell, *Why I Am Not a Christian*

J. L. Mackie, *The Miracle of Theism*

Graham Oppy, *Arguing About Gods*

Stanford Encyclopedia of Philosophy, "Cosmological Argument"

Chapter 8: Who Created God?

The Child's Question That Adults Avoid

A child asks a simple question.

"Who created God?"

The adults may smile. Some may become uncomfortable. Some may say, "God was not created." Some may say, "God always existed." Some may say, "That is a mystery." Some may say, "You should not ask that." Some may say, "God is beyond human understanding."

But the child's question is not foolish.

It is one of the most serious questions in the entire debate about God.

If the universe needs a cause, why does God not need one?

If everything must be explained, why is God allowed to remain unexplained?

If the universe cannot simply exist, why can God simply exist?

If the universe cannot be eternal, why can God be eternal?

If the universe cannot be the final reality, why can God be the final reality?

This question is powerful because it exposes a hidden rule in many religious arguments.

The believer often says that everything needs an explanation, but then exempts God from that demand. The universe must explain itself, but God does not. Matter cannot be eternal, but God can. The cosmos cannot be necessary, but God can. Physical reality cannot be the stopping point, but God can be.

This may satisfy someone already committed to belief.

But to the questioning mind, it creates a serious problem.

The explanation has been protected, not proven.

The mystery has not been solved.

It has been moved behind the word God.

This chapter asks whether God really explains the universe, or whether God simply becomes a larger mystery placed behind the universe.

The Religious Stopping Point

Every worldview needs a stopping point.

At some point, explanation reaches a boundary. The believer may stop with God. The atheist may stop with reality, physical law, the universe, the multiverse, quantum fields, or some deeper natural foundation not yet fully understood.

The question is not whether we need a stopping point.

The question is whether God is a better stopping point than nature itself.

The religious answer is usually this:

God is eternal.

God is uncaused.

God is necessary.

God is self-existent.

God does not depend on anything else.

God is not part of the universe.

God is the reason the universe exists.

This answer has emotional and philosophical force. It gives the mind a final resting place. It prevents an infinite chain of causes. It says that reality is grounded in a supreme being.

But we must ask:

How do we know this?

How do we know that God is necessary?

How do we know that God is uncaused?

How do we know that God exists outside time?

How do we know that God is the final explanation rather than a human invention placed at the end of our questions?

The believer may say, “Because the universe needs an explanation.”

But that does not prove God.

It only shows that the believer wants an explanation.

Wanting an explanation is not the same as having one.

Serious Theology and Necessary Being

A fair discussion must acknowledge something important.

Serious theologians do not usually imagine God as one more physical object inside the universe. They do not think God is like a large invisible person floating somewhere in space. They often describe God as necessary being, the ground of existence, the source of all contingent reality, or the reason anything exists at all.

This is a more sophisticated idea than the simple childhood image of God above the clouds.

It deserves to be taken seriously.

But taking it seriously does not mean accepting it as true.

Calling God necessary does not prove that God exists.

Calling God the ground of existence does not show that such a being is real.

Saying that God is beyond space and time does not explain how God creates space and time.

Saying that God is the source of all reality does not tell us how this is known.

Even if some necessary foundation of reality exists, it does not follow that this foundation is personal, conscious, loving, moral, prayer-answering, or connected to any religion.

This is where religious argument often moves too quickly.

It begins with a deep philosophical idea: perhaps reality has a necessary foundation.

Then it gives that foundation the name God.

Then it quietly adds the features of religion: mind, will, purpose, love, command, judgment, worship, scripture, prayer, heaven, and moral authority.

But those additions are not proven.

They are imported.

A necessary foundation, if it exists, is not automatically the God people worship.

The Problem of Special Pleading

Special pleading occurs when someone applies a rule to everything else but makes an exception for the thing he wants to defend.

In many arguments for God, the rule is:

Everything needs a cause.

Then the exception is:

Except God.

Or the rule is:

Everything needs an explanation.

Then the exception is:

Except God.

Or the rule is:

Complex reality cannot simply exist.

Then the exception is:

Except the most complex reality imaginable, God.

This is the problem.

If the believer says everything needs a cause, then God needs a cause.

If the believer says not everything needs a cause, then the universe may not need a cause.

Either way, the simple argument weakens.

The believer may respond that God is not like ordinary things. God is necessary, eternal, and uncaused. But this must be argued, not merely asserted.

Why should we accept God as the exception?

Why not accept the deepest natural reality as the exception?

Why not say that the universe, or some underlying physical reality, is necessary or uncaused?

The religious answer often gives God special status at the beginning. But that is exactly what needs to be proven.

Science-based atheism does not say it has solved every ultimate mystery. It says we should not protect one mystery from questioning while demanding explanations from everything else.

That is not honest reasoning.

That is special pleading.

Is God Simpler Than the Universe?

Many believers think God explains the universe because God is the ultimate source of everything.

But is God simpler than the universe?

Consider what traditional religion often claims about God.

God is timeless.

God is spaceless.

God is immaterial.

God is all-powerful.

God is all-knowing.

God is perfectly good.

God is conscious.

God has intention.

God creates matter, energy, space, time, and law.

God hears prayer.

God knows every thought.

God judges human actions.

God may perform miracles.

God may grant eternal life.

God may punish or reward.

This is not a simple explanation.

This is an enormous claim.

In ordinary reasoning, we usually prefer simpler explanations over more complicated ones, especially when the more complicated explanation has no independent evidence. If we are trying to explain the universe, and we introduce an invisible, timeless, spaceless, conscious, all-powerful being, we have not made the mystery smaller.

We have made it larger.

The universe is mysterious.

But God is more mysterious.

If matter, energy, space, time, and law are difficult to explain, how much more difficult is an infinite supernatural mind that creates them all?

God does not simplify the problem.

God multiplies the problem.

This is one reason science-based atheism is intellectually stronger. It does not add an even greater mystery to explain the first mystery.

It begins with reality and investigates from there.

The Problem of an Uncaused Mind

The traditional God is not merely an uncaused force.

God is usually described as mind.

A cosmic mind.

A supreme consciousness.

A being with intention, knowledge, will, purpose, and power.

But minds are complex.

Every mind we actually know is connected to a brain or nervous system. Human minds depend on memory, perception, emotion, attention, development, language, and experience. Animal minds also arise from living bodies. We have no clear evidence of a mind existing without physical structure.

Religion asks us to accept the most powerful mind imaginable without a brain, without a body, without time, without development, without environment, without learning, and without experience in any ordinary sense.

That is a large claim.

It is not made more reasonable by calling it necessary.

An uncaused mind is not simpler than an unexplained universe.

It is more difficult.

The universe contains mind at least in some places, such as human beings. But the minds we know appear through natural biological processes. The religious claim reverses this. It says a mind exists before biology, before matter, before space, before time, before anything physical.

Where is the evidence for that?

There is none.

God as an uncaused mind is not an explanation supported by experience. It is a projection of human mentality onto ultimate reality.

This is exactly the pattern seen throughout this series.

Human beings explain the unknown by imagining a person behind it.

But the universe does not need a person behind it.

Necessary Being Does Not Mean Religious God

Some philosophers avoid the phrase "Who created God?" by arguing that God is a necessary being.

A necessary being is not something that happens to exist. It cannot fail to exist. It exists by its own nature. It does not depend on anything else.

This is a more sophisticated argument than simply saying God was always there.

But it still faces a major problem.

Calling something necessary does not prove that it exists.

One cannot define God into existence by saying, “God is the being that must exist.” The key question remains: does such a being actually exist?

Also, why must necessary reality be a person?

Why must necessary reality be conscious?

Why must necessary reality be morally perfect?

Why must necessary reality care about human beings?

Why must necessary reality be the God of any religion?

If there is a necessary foundation of existence, it does not automatically follow that this foundation is God in the religious sense. It could be something far stranger, deeper, and more natural than theology imagines.

It could be physical law.

It could be quantum reality.

It could be a deeper structure beyond current physics.

It could be something we have not yet conceptualized.

Or perhaps existence itself is the brute fact.

The religious move from “there may be necessary reality” to “therefore God” is too quick.

Necessary being is a philosophical possibility.

It is not proof of a divine person.

Brute Facts and the Desire for Final Answers

A brute fact is something that exists without a deeper explanation, or at least without one available in the framework being used.

Many people dislike brute facts. The mind wants explanation. It wants reason. It wants a final answer. This desire is natural.

But reality is not obligated to satisfy our desire for final explanation.

The believer often says, "The universe cannot be a brute fact."

But then the believer treats God as a brute fact.

God just exists.

God needs no cause.

God needs no explanation.

God is the final answer.

The atheist asks: why is that allowed for God but not for reality itself?

If something must be accepted as ultimate, why not accept the natural world, or some deeper natural foundation, rather than a supernatural person?

This does not mean the universe as we observe it is the final explanation. Science may discover deeper levels of reality. Our current cosmology may be incomplete. Physics may change. Quantum gravity may transform our understanding. Multiverse ideas may or may not prove useful.

But the direction of explanation should remain natural until evidence requires otherwise.

Science has earned trust by explaining nature naturally.

Religion has repeatedly filled gaps with God, then retreated when science explained more.

So when we reach a deep mystery, history gives us a warning:

Do not insert God too quickly.

The unknown has never been safe territory for religion.

The Universe as the Stopping Point

Suppose someone says, “God just exists.”

The atheist can say, “Perhaps reality just exists.”

Suppose someone says, “God is eternal.”

The atheist can say, “Perhaps the deepest natural reality is eternal, even if our observable universe has a finite past.”

Suppose someone says, “God is necessary.”

The atheist can say, “Perhaps physical reality, or the foundation behind physical reality, is necessary.”

Suppose someone says, “God is beyond time.”

The atheist can say, “Perhaps the deepest level of reality is not temporal in the ordinary way.”

The believer may object, “But that sounds mysterious.”

Yes.

It is mysterious.

But God is also mysterious.

The question is not whether mystery remains. The question is which position is more disciplined.

Science-based atheism does not pretend to have final answers beyond evidence. It says that if a stopping point is needed, we should not choose a supernatural mind without evidence. We should continue to investigate reality itself.

That is a more honest path.

The universe, or whatever deeper natural reality underlies it, may be the ultimate stopping point.

We may not yet know.

But not knowing does not make God the answer.

Does God Explain Anything Better?

A useful explanation should do something.

It should clarify.

It should reduce confusion.

It should connect facts.

It should make predictions.

It should help us understand why reality is this way rather than another way.

So we must ask: does God do this?

If God created the universe, why this universe?

Why so vast?

Why so old?

Why mostly empty?

Why mostly hostile to life?

Why billions of galaxies before human beings appeared?

Why laws rather than direct divine action?

Why suffering?

Why extinction?

Why cosmic violence?

Why dark matter and dark energy?

Why a universe that looks natural whether or not God exists?

The believer may answer, "God's reasons are mysterious."

But that is not an explanation.

That is the end of explanation.

If every difficult question is answered with "God's mystery," then God explains nothing clearly. A small universe would be God's will. A vast universe is God's will. A universe with life is God's will. A universe without life would also be God's will. A peaceful universe would reflect God's goodness. A violent universe would reflect God's mysterious plan.

An explanation that fits every possible outcome explains no specific outcome.

Science is different.

It explains why some outcomes follow from certain conditions. It can be wrong. It can be corrected. But it gives us a real path.

God does not explain the universe better.

God makes the universe depend on an even greater mystery.

"God Is Beyond Human Understanding"

Another common answer is that God is beyond human understanding.

This may be emotionally satisfying, but it creates a problem.

If God is beyond human understanding, then how do believers know so much about God?

How do they know God is good?

How do they know God is personal?

How do they know God wants worship?

How do they know God created the universe?

How do they know God listens to prayer?

How do they know God cares about human beings?

How do they know God is not indifferent, limited, unconscious, or nothing like religion imagines?

The phrase "God is beyond understanding" is often used selectively.

When God is questioned, God becomes mysterious.

When God is worshiped, God becomes knowable.

When contradictions appear, God is beyond us.

When doctrines are taught, God is clear enough.

This is not consistent.

Science-based atheism asks for evidence. If God is beyond understanding and beyond testing, then God cannot be used as a clear explanation. A being beyond explanation cannot explain the universe.

This is a key point.

If God is too mysterious to be understood, then God is too mysterious to function as an explanation.

Mystery cannot solve mystery by becoming a larger mystery.

The Creator Creates More Questions

If God created the universe, many new questions arise.

Why did God create at all?

Was God lacking something before creation?

If God was perfect, why create anything?

If God knew everything that would happen, why create a universe with suffering?

Why create a universe through billions of years of cosmic violence, stellar death, extinction, disease, and struggle?

Why create human beings so late in cosmic history?

Why create a universe in which God's existence is so unclear?

Why create minds capable of honest disbelief?

Why reveal truth through ancient cultures rather than universal evidence?

Why make the universe look exactly as if it operates by natural law?

These questions do not disappear by saying God is mysterious.

They grow.

The creator hypothesis is often presented as if it solves the ultimate problem. But in reality, it creates many additional problems.

A natural universe may be mysterious, but it is at least the thing we observe.

A supernatural creator is not observed.

It is inferred, imagined, defended, and protected from ordinary standards of evidence.

Science-based atheism asks a fair question:

Why add this invisible layer?

If the universe can be investigated naturally, and if God adds no clear explanatory power, then God is unnecessary.

The Difference Between Explanation and Assertion

An assertion is a statement.

An explanation is a statement that clarifies.

“God created the universe” is an assertion.

But what does it clarify?

If it does not tell us how God creates, why God creates, what evidence shows God created, what predictions follow, or how the claim could be tested, then it is not a strong explanation.

It is a declaration.

Religious language often sounds explanatory because it has familiar grammar.

God made the universe.

God designed the laws.

God chose the constants.

God created time.

God sustains existence.

But grammar is not evidence.

A sentence can be grammatically complete and scientifically empty.

"An invisible cosmic mind created space-time by timeless will" is a sentence. But it does not explain in the way physics explains. It does not provide a model. It does not connect to measurable evidence. It does not predict observations better than natural cosmology.

This is why science-based atheism is stronger.

It does not accept impressive words as explanation.

It asks what the words actually do.

In the case of God, they usually do very little.

They stop questioning.

They do not answer it.

Why the Child's Question Still Matters

The child's question remains.

Who created God?

The adult answer may be, "No one. God is uncreated."

But then the atheist may ask, "Why can reality not be uncreated?"

The adult answer may be, "God is eternal."

But then the atheist may ask, "Why can the deepest natural reality not be eternal?"

The adult answer may be, "God is necessary."

But then the atheist may ask, "Why must necessary reality be God?"

The adult answer may be, "God is beyond understanding."

But then the atheist may ask, "Then how does God explain anything?"

This is not childish rebellion.

It is clear reasoning.

The question "Who created God?" exposes the special exception at the heart of many religious arguments. It forces us to ask whether God is truly an explanation or merely a protected mystery.

In a science-based worldview, we do not protect ideas from questioning because they are sacred.

We question them because truth matters.

God should not receive immunity from the reasoning used against the universe.

If the universe needs explanation, so does God.

If God does not need explanation, perhaps the universe does not need that kind of explanation either.

That is the force of the question.

The Strongest Religious Reply

A thoughtful believer may say:

"God does not need a creator because God is not a created thing. God is necessary being itself. Everything in the universe is contingent, changing, and dependent. God is different. God is the ground of existence, not one object among others. Asking who created God misunderstands what God means."

This is the strongest religious reply.

It deserves respect.

It correctly says that serious theology does not imagine God as merely one more object inside the universe. It also correctly notes that the question of ultimate existence may require deeper philosophical language.

But the reply still has problems.

First, saying God is necessary does not prove God exists.

Second, saying God is the ground of existence does not explain how we know this.

Third, this answer does not establish a personal God.

Fourth, it does not prove any religion.

Fifth, it does not explain how a timeless, spaceless, immaterial ground becomes a conscious creator.

Sixth, it gives no testable evidence.

Seventh, it may simply move the mystery from the universe to God.

The believer may define God as the ultimate explanation. But definition is not demonstration.

Science-based atheism remains more careful.

It says: perhaps reality has a deepest foundation, but we do not have good evidence that this foundation is a supernatural person.

That is the more honest position.

Evidence-Based Response

The evidence-based response is simple.

God does not solve the mystery of existence.

God relocates it.

If everything needs a cause, God needs a cause.

If God does not need a cause, then not everything needs a cause.

If God can be eternal, perhaps the deepest natural reality can be eternal.

If God can be necessary, perhaps natural reality can be necessary.

If God is beyond explanation, then God cannot function as a clear explanation.

The claim that God created the universe adds no scientific mechanism, no prediction, no measurement, and no testable model. It does not explain why this universe exists rather than another. It does not identify which God exists. It does not establish prayer, scripture, miracle, heaven, or moral authority.

It simply puts a divine name at the place where knowledge becomes difficult.

Science-based atheism does not pretend that the ultimate nature of reality is fully understood. It accepts mystery where mystery remains. But it refuses to call that mystery God without evidence.

That is not arrogance.

It is intellectual honesty.

Chapter Summary

The question “Who created God?” is often treated as a childish question, but it is philosophically serious. It exposes a key weakness in many creator arguments.

If everything needs a cause, then God needs a cause. If God does not need a cause, then it is false that everything needs a cause. If God can be uncaused, eternal, or necessary, then the deepest natural reality may also be uncaused, eternal, or necessary.

Serious theology may describe God as necessary being or the ground of existence. But this does not prove that God exists. It does not prove that ultimate reality is personal, conscious, loving, moral, or connected to any religion.

God does not remove the mystery of existence.

God often becomes a larger mystery placed behind the first mystery.

Science-based atheism accepts that deep questions remain, but it does not turn those questions into God.

Key Points

The question “Who created God?” is serious, not childish.

Many religious arguments apply rules to the universe but exempt God.

This is often special pleading.

Serious theology may call God necessary being, but that does not prove God exists.

A necessary foundation of reality, if one exists, is not automatically personal or divine.

God is not simpler than the universe.

An uncaused cosmic mind is a very large claim with no clear evidence.

If God can be uncaused, perhaps the deepest natural reality can be uncaused.

God often relocates mystery rather than explaining it.

Science-based atheism accepts mystery without turning it into theology.

Transition to the Next Chapter

The first-cause argument does not prove God.

The question "Who created God?" exposes serious weaknesses in the creator explanation.

But another argument remains powerful for many modern believers. It does not begin with cause. It begins with apparent precision.

The universe seems to have certain constants, laws, and initial conditions that allow stars, atoms, chemistry, planets, and life. Change them, and life as we know it might not exist.

Does this fine-tuning prove a cosmic designer?

In the next chapter, we will examine fine-tuning, perhaps the strongest modern design argument for God.

Sources

David Hume, *Dialogues Concerning Natural Religion*

Bertrand Russell, *Why I Am Not a Christian*

J. L. Mackie, *The Miracle of Theism*

Graham Oppy, *Arguing About Gods*

Richard Swinburne, *The Existence of God*

William Lane Craig, writings on the Kalam cosmological argument

Gottfried Wilhelm Leibniz, writings on necessary being and sufficient reason

Stanford Encyclopedia of Philosophy, "Necessary and Contingent Truths"

Chapter 9: Fine-Tuning: The Strongest Modern Design Argument

The Universe That Seems Too Precise

Imagine a hidden control room behind reality.

On one wall are dials. One controls gravity. Another controls electromagnetism. Another controls the forces inside atoms. Another controls the expansion of the universe. Another controls the masses of particles. Another controls the conditions that allow stars, chemistry, planets, and life.

Now imagine turning one dial slightly.

Stars never form.

Turn another.

Atoms cannot hold together.

Turn another.

The universe expands too fast for galaxies to form.

Turn another.

The universe collapses too quickly.

Turn another.

Chemistry becomes impossible.

Turn another.

Life never appears.

This image is not exact physics, but it captures why the fine-tuning argument feels so powerful.

The universe seems to have features that allow complexity. Stars exist. Atoms exist. Carbon exists. Chemistry exists. Planets exist. Life exists. Conscious beings exist. We are here, asking why.

To many believers, this does not look accidental.

It looks arranged.

They say the universe appears delicately balanced for life. If the constants of nature had been even slightly different, life as we know it might not have existed. Therefore, they argue, the universe must have been designed. Fine-tuning requires a fine-tuner. Cosmic order requires a cosmic mind. The laws and constants of physics point beyond physics to God.

This is one of the strongest modern arguments for God.

It deserves serious attention.

Older religious arguments often depended on ignorance that science later corrected. Thunder did not need a god.

Disease did not need demons. Species did not need special creation. Planetary motion did not need angels.

Fine-tuning is more subtle.

It does not deny science. It uses science. It accepts physics, cosmology, and mathematics, then asks why the universe has the kind of laws and constants that permit complexity and life.

That is why this argument is attractive to thoughtful believers.

But attractive does not mean true.

Fine-tuning is a real question.

It is not proof of God.

What Fine-Tuning Means

Fine-tuning does not mean that the universe was literally adjusted by a divine hand.

At least, that is not what the scientific issue means.

Fine-tuning means that certain features of the universe appear to fall within ranges that allow complex structures such as atoms, stars, galaxies, planets, chemistry, and life. If some values were very different, the universe might have been sterile, chaotic, short-lived, or unable to produce complexity.

Examples often discussed include the strength of gravity, the strength of electromagnetism, the strong and weak nuclear forces, the cosmological constant, the masses of particles, and the initial conditions of the early universe.

The key point is this:

The universe seems life-permitting.

Not necessarily life-centered.

Not necessarily human-centered.

Life-permitting.

This distinction matters.

The fine-tuning argument often adds more than the evidence allows. It begins with the observation that the universe permits life, then quickly concludes that the universe was designed for life, and often even for human beings.

But those are different claims.

A universe that permits life is not automatically a universe designed for life.

A pond permits frogs to live. That does not prove the pond was designed for frogs. A cave permits bats to live. That does not prove the cave was designed for bats. A planet permits organisms to evolve. That does not prove the planet was made for those organisms.

Life fits the conditions of this universe because life arose within this universe.

Still, the deeper question remains.

Why does the universe have life-permitting conditions at all?

That is a serious question.

Science should not dismiss it casually.

But religion should not claim victory too quickly.

The Emotional Power of Design

The fine-tuning argument feels persuasive because human beings are used to design.

When we see a watch, we infer a watchmaker.

When we see a book, we infer an author.

When we see a house, we infer a builder.

When we see a machine, we infer an engineer.

This habit is reasonable in ordinary life. Complex functional objects usually come from minds. A heart monitor does not assemble itself in a hospital room. A bridge does not appear without engineering. A smartphone is not produced by wind and rain.

So when people look at the universe and see order, law, mathematics, structure, and life, they naturally imagine a cosmic designer.

This is emotionally understandable.

But the universe is not a watch.

It is not a machine sitting inside a larger workshop.

It is not one object among many comparable objects whose origin we can inspect.

We know watches are designed because we have independent experience of watchmakers. We know books are written because we know authors. We know houses are built because we know builders.

But we do not have independent evidence of universe-makers.

We have only one observable universe. We do not know the full range of possible universes. We do not know whether the constants could have been different. We do not know whether deeper physical laws determine them. We do not know whether many cosmic domains exist. We do not know the probability distribution of possible laws and constants.

So the analogy to human design is weak.

It feels strong because the mind likes familiar patterns.

But cosmic origin may not follow the patterns of carpentry, engineering, or clockmaking.

The universe is not a human artifact.

Carbon, Stars, and the Temptation of Purpose

One of the most moving examples in fine-tuning discussions involves carbon.

Carbon is essential for life as we know it. It can form complex molecules. It can build long chains. It can combine in many ways. Without carbon chemistry, life as we know it would not exist.

But carbon did not simply appear.

It was made in stars.

Stars formed from clouds of gas. Inside stars, nuclear reactions built heavier elements. Some stars died violently and scattered those elements into space. Later, that material became part of new stars, planets, oceans, rocks, and living bodies.

The carbon in our cells has a cosmic history.

This is a beautiful thought.

It is also a dangerous point if handled carelessly.

A believer may say, "Surely this was planned. Stars made carbon, carbon allowed life, life produced consciousness, and consciousness now looks back at the stars. How can this not be design?"

The emotional pull is real.

But the conclusion does not follow.

The fact that life depends on carbon does not prove that carbon was made for life. The fact that stars produce heavy elements does not prove that stars intended life. The fact that human beings are made of cosmic material does not prove that the cosmos was aiming at human beings.

Science gives us a natural chain:

Stars.

Elements.

Planets.

Chemistry.

Life.

Consciousness.

Religion adds intention to that chain.

But the intention is not shown by the evidence.

The natural story is already magnificent.

It does not become more true by adding God.

Life-Permitting Is Not Human-Centered

Fine-tuning arguments often sound as if the universe was carefully prepared for us.

But the universe is not comfortable for us.

Most of it is deadly.

Space has no air.

Stars burn.

Radiation kills.

Black holes destroy.

Most planets are lifeless.

Many worlds are frozen, scorched, toxic, crushed, or airless.

Even Earth, our only known home, is safe only within narrow limits. We survive in a thin layer of atmosphere, within a small range of temperature and pressure, protected from much of the universe by distance, magnetic field, atmosphere, and biological adaptation.

If the universe was designed for human beings, the design is strange.

Almost all of it is unusable by human beings.

The believer may say, “But the universe had to be large and old to produce the conditions for life.”

That may be true.

Stars had to form. Heavy elements had to be produced. Planets had to develop. Time was needed.

But this still does not prove design.

It only shows that life depends on cosmic conditions.

Science-based atheism can fully accept that. Life is part of the universe. We are made of atoms created through cosmic history. We depend on stars, chemistry, gravity, and time.

But dependence is not evidence of divine intention.

A fish depends on water.

That does not mean the ocean was designed for that fish.

Life arises where conditions permit it.

That is not the same as saying the universe was made for life.

The Probability Problem

Many fine-tuning arguments use probability.

They say the chance of a life-permitting universe is extremely small. Therefore, design is the best explanation.

This sounds impressive.

But there is a serious problem.

To calculate probability, we need a meaningful range of possibilities and a probability distribution. In simple language, we need to know what values were possible and how likely each value was.

With dice, probability works because we know the sides. If we roll a fair six-sided die, we can say the chance of rolling a four is one in six.

With cards, probability works because we know the deck.

With coins, probability works because we know the possible outcomes.

But with the constants of nature, we do not know the cosmic deck.

We do not know whether the constants could have taken any possible value.

We do not know whether some values are impossible.

We do not know whether deeper laws determine them.

We do not know whether they vary across a larger multiverse.

We do not know whether life of some different kind could exist under conditions we have not imagined.

So when someone says, "The probability of this universe is incredibly small," we must ask:

Small compared with what?

Compared with which possible universes?

With what probability distribution?

Based on what deeper theory?

Without this, the probability claim becomes less secure than it sounds.

It may be true that life-permitting conditions are special.

But turning that into a precise probability for God is not justified.

Fine-tuning may show that we have a question.

It does not show that we have calculated the answer.

Chance Is Not the Only Alternative to God

Religious arguments often present a false choice:

Either blind chance or God.

This is misleading.

The real possibilities are broader.

Perhaps the constants are determined by deeper physical laws.

Perhaps they are connected in ways we do not yet understand.

Perhaps some values that seem possible are not physically possible.

Perhaps there are many cosmic domains with different conditions.

Perhaps observer selection explains why we find ourselves in a life-permitting region.

Perhaps life can exist under conditions very different from what we currently imagine.

Perhaps some facts about the universe are brute facts.

Perhaps future physics will reveal explanations that no one has yet imagined.

The point is not that any one of these answers is definitely correct.

The point is that God is not the only option.

This matters deeply.

The fine-tuning argument often gains power by narrowing the reader's imagination. It says: chance is absurd, therefore design is obvious.

But science is not limited to chance.

Science looks for necessity, law, structure, selection, deeper theory, and evidence.

The proper answer to fine-tuning is not automatic atheistic certainty.

It is also not automatic God.

The proper answer is continued inquiry.

Science-based atheism remains stronger because it does not rush from difficulty to deity.

The Problem of Unknown Physics

Fine-tuning arguments often assume that the constants of nature are free to vary independently.

But what if they are not?

What if a deeper physical theory explains why they have the values they have?

What if the values are connected?

What if only certain combinations are possible?

What if the universe could not have been otherwise in the way the argument assumes?

We do not yet know.

This is not a weakness of science.

It is honesty.

Physics is not complete. We do not yet have a final theory that unifies all forces and explains all constants from deeper principles. Quantum gravity remains unfinished. The foundations of cosmology remain under investigation.

But the history of science teaches caution.

Again and again, things that once seemed mysterious later received natural explanation.

Lightning became electricity.

Disease became microbiology.

Planetary motion became gravity.

Design in life became evolution.

Matter became chemistry.

Mind became increasingly linked to the brain.

So when someone says, "We do not know why these constants have these values, therefore God," we should remember the past.

The unknown is not evidence for God.

It is an invitation to continue inquiry.

Perhaps future physics will explain some or all apparent fine-tuning. Perhaps it will not. But lack of current explanation does not justify supernatural certainty.

Science-based atheism is comfortable saying:

We do not yet know.

Religion often wants to say:

We do not know, therefore God.

The first is honest.

The second is premature.

The Anthropic Clue

The anthropic principle will be discussed more fully in the next chapter, but it belongs briefly here.

We observe a universe compatible with our existence because if the universe were not compatible with our existence, we would not be here to observe it.

This sounds simple.

But it matters.

Any observer will necessarily find itself in conditions where observers can exist. That does not prove the conditions were designed for the observer. It only shows that observation is possible only where observers are possible.

A person born on Earth should not be surprised that Earth has conditions compatible with human life. If Earth did not have such conditions, that person would not exist.

Similarly, we should not be shocked that the universe we observe permits observers.

We could not observe any other kind.

This does not solve every fine-tuning question. It does not explain why there is a life-permitting universe at all.

But it weakens the claim that our existence is surprising evidence of design.

We are not independent spectators looking at a universe chosen at random.

We are products of the universe we observe.

That changes the interpretation.

The universe may not be fitted to us.

We may be fitted to the universe.

Fine-Tuning and the God of the Gaps

Fine-tuning can become a sophisticated version of the God of the gaps.

The argument says:

We do not know why the constants have these values.

We do not know why the laws permit life.

We do not know why the early universe had the conditions it had.

Therefore, God.

But the word therefore is doing too much work.

A gap in knowledge does not become evidence for God simply because the gap is large or difficult.

The God of the gaps has always been tempting because human beings dislike uncertainty. We want closure. We want final answers. We want mystery to become meaning.

But history has not been kind to gap-based religion.

God was placed in the storm.

Then meteorology advanced.

God was placed in disease.

Then germ theory advanced.

God was placed in species.

Then evolution advanced.

God was placed in the motions of heaven.

Then physics advanced.

Now God is placed in fine-tuning.

The pattern is familiar.

Science-based atheism says: wait. Investigate. Do not worship the gap. Do not call ignorance a creator. Do not place God where evidence has not yet reached.

Fine-tuning may be a deep scientific and philosophical problem.

It is not a divine signature.

The Problem of a Designed Designer

If fine-tuning requires a fine-tuner, then we must ask about the fine-tuner.

A God capable of designing the universe would be astonishingly complex.

This God would understand all possible laws, constants, universes, outcomes, particles, fields, stars, galaxies, chemistry, life, consciousness, and future history. This God would choose one universe rather than another. This God would have intention, knowledge, power, and purpose beyond imagination.

How is this a simpler explanation?

A designer capable of fine-tuning the universe seems even more in need of explanation than the universe itself.

The believer may say God is necessary or simple in a theological sense. But this does not answer the problem. A conscious being with infinite knowledge and creative power is not simple in any ordinary explanatory sense.

God is introduced to explain fine-tuning, but God appears more mysterious than the fine-tuning itself.

Where did God's mind come from?

Why does God have the power to create?

Why does God know how to design universes?

Why does God want life?

Why does God create this universe rather than another?

Why does God create through laws that make God unnecessary in practice?

Again, the mystery is not solved.

It is moved upward.

Science-based atheism does not make that move. It seeks natural explanation without adding an even larger unexplained mind behind reality.

The Universe Does Not Look Perfectly Designed

If the universe were designed for life, we might expect it to be more life-friendly.

Instead, the universe is overwhelmingly empty and hostile.

Even life on Earth is not protected from suffering. Asteroids strike. Species go extinct. Diseases spread. Genetic errors occur. Natural disasters kill. Organisms survive by consuming one another. Most species that ever lived are gone.

Fine-tuning arguments often focus on the fact that life is possible.

They rarely focus enough on the fact that life is difficult, fragile, rare, and surrounded by death.

A universe designed by a loving God for life should perhaps look more life-centered than this.

But a natural universe explains the evidence well.

Life appears where conditions allow it.

It struggles.

It adapts.

It persists for a time.

It disappears when conditions fail.

There is no need to imagine that the entire cosmos was lovingly designed for human beings.

In fact, the universe looks exactly like what science would expect: vast physical processes producing complexity in limited regions under suitable conditions.

That is enough.

The design argument asks us to see intention where natural process is sufficient.

Beauty Is Not Evidence

Many people feel that the universe is too beautiful to be accidental.

They see stars, galaxies, mathematics, symmetry, and order, and they feel that beauty points to God.

This feeling is understandable.

The universe is beautiful.

But beauty is not evidence for design.

A snowflake is beautiful because of natural processes.

A rainbow is beautiful because of light and water.

A crystal is beautiful because of molecular structure.

A galaxy is beautiful because of gravity, stars, gas, dust, and motion.

Beauty can arise naturally.

Human beings also bring beauty to the universe through perception. We see patterns. We feel awe. We experience symmetry, scale, light, and motion emotionally. The beauty is partly in the world and partly in the mind that sees the world.

This does not make beauty unreal.

It makes it natural.

Science-based atheism does not deny beauty. It refuses to confuse beauty with proof of God.

The fact that the universe moves us emotionally does not show that a divine mind made it.

It shows that conscious beings can experience awe.

That is wonderful enough.

Fine-Tuning and Religious Overreach

Even if fine-tuning suggested design, it still would not prove any specific religion.

It would not prove Christianity.

It would not prove Islam.

It would not prove Hinduism.

It would not prove Judaism.

It would not prove any scripture.

It would not prove miracles.

It would not prove prayer.

It would not prove heaven or hell.

It would not prove that God cares about human morality.

It would not prove that God created human beings in a special spiritual sense.

At most, fine-tuning might suggest some kind of cosmic designer.

But religious believers usually want much more than that.

They want the God they already worship.

This is a major problem.

The fine-tuning argument begins with physics and ends with religion, but the bridge is weak. Even if a designer were inferred, the designer could be deistic, indifferent, limited, multiple, impersonal, alien, or unlike any religious God.

This does not mean such alternatives are likely.

It means the argument does not get believers where they want to go.

Science-based atheism can say clearly:

Fine-tuning does not prove God.

And it certainly does not prove any particular religion.

A Better Explanation: We Belong to This Universe

There is another way to think about fine-tuning.

We are not strangers dropped into the universe.

We are products of it.

The atoms in our bodies come from stars. The chemistry of life depends on cosmic history. Earth formed from earlier material. Life evolved under local conditions. Human minds emerged from biological evolution. We are not separate from the universe, looking at it from outside. We are inside the universe, made from its matter and shaped by its laws.

This changes the emotional interpretation.

Instead of saying, "The universe was made for us," we can say, "We arose from the universe."

That is more humble.

It is also more scientific.

We fit the universe because we came from it.

Life is shaped by the conditions that permit it. Consciousness appears where biology evolves nervous systems capable of experience. Human beings ask cosmic questions because matter, under certain conditions, became living, thinking matter.

This does not require God.

It requires nature.

And nature is enough.

The Strongest Religious Reply

A thoughtful believer may say:

"Fine-tuning is too precise to be dismissed. The constants and laws of nature appear delicately balanced for life. Chance alone is not a reasonable explanation. The best explanation is that an intelligent creator designed the universe so that life could exist. The multiverse is speculative, and deeper laws are unknown. God is a better explanation than blind chance."

This reply is serious.

It correctly says that fine-tuning should not be dismissed. It correctly challenges simplistic appeals to chance. It correctly notes that multiverse ideas are not proven and that physics is incomplete.

But the reply still fails as proof.

First, the choice is not only chance or God.

Second, we do not know whether the constants could vary freely.

Third, we do not know the probability distribution of possible universes.

Fourth, we do not know whether deeper laws determine the values.

Fifth, observer selection effects matter.

Sixth, God is not a simple explanation.

Seventh, God gives no testable mechanism.

Eighth, fine-tuning does not prove a personal religious God.

Fine-tuning is a genuine mystery.

But a genuine mystery is still not proof of God.

Evidence-Based Response

The evidence-based response is clear.

Fine-tuning is one of the strongest modern arguments for God because it takes science seriously and asks a real question. The universe does appear to have features that allow complexity, chemistry, stars, planets, and life.

But fine-tuning does not prove God.

The probabilities are not known. The possible range of constants is not known. Deeper physical laws may explain more than we now understand. Observer selection effects must be considered. Multiverse ideas, while unproven, show that design is not the only possible answer. The universe is life-permitting only in limited regions and overwhelmingly hostile to life elsewhere.

Most importantly, God does not explain fine-tuning scientifically.

God gives no mechanism, no measurable model, no testable prediction, and no way to distinguish one religious claim from another.

Fine-tuning is a question for physics, cosmology, and philosophy.

It is not a proof of God.

Science-based atheism can accept the mystery without surrendering to divine design.

Chapter Summary

Fine-tuning is one of the most serious modern arguments for God. It begins with the observation that certain features of the universe appear to allow stars, atoms, chemistry, planets, and life. If these features were very different, life as we know it might not exist.

This raises a real question.

But it does not prove God.

We do not know the full range of possible constants. We do not know whether the constants could vary freely. We do not know whether deeper laws determine them. We do not know whether many cosmic domains exist. We do not know whether other kinds of life are possible under different conditions.

The universe is life-permitting, but not obviously life-centered. Most of it is hostile to life. Fine-tuning may show that we have much to understand, but it does not show that a divine designer exists.

God is not required.

Key Points

Fine-tuning means that certain features of the universe appear to allow complexity and life.

Life-permitting does not mean human-centered.

The fine-tuning argument feels powerful because humans naturally infer design from order.

The universe is not a watch, a house, or a machine made by a known designer.

Probability claims about fine-tuning are difficult because we do not know the full range of possible universes.

Chance is not the only alternative to God.

Deeper physical laws may explain what currently appears finely tuned.

The anthropic principle weakens simple design arguments.

The multiverse is not proven, but it shows that design is not the only possible explanation.

God does not explain fine-tuning scientifically.

Fine-tuning is a mystery, not proof.

Transition to the Next Chapter

Fine-tuning leads naturally to another idea.

We observe a universe compatible with life because only such a universe could contain observers like us. This is called the anthropic principle.

It is often misunderstood. It does not solve every mystery. It does not prove atheism by itself. But it does weaken the claim that our existence is surprising evidence of divine design.

In the next chapter, we will examine the anthropic principle and ask why we find ourselves in a universe where observers can exist.

Sources

Martin Rees, *Just Six Numbers*

Victor Stenger, *God: The Failed Hypothesis*

Sean Carroll, *The Big Picture*

Stephen Hawking and Leonard Mlodinow, *The Grand Design*

Brian Greene, *The Fabric of the Cosmos*

David Hume, *Dialogues Concerning Natural Religion*

Stanford Encyclopedia of Philosophy, "Fine-Tuning"

Stanford Encyclopedia of Philosophy, "Cosmology and Theology"

Chapter 10: The Anthropic Principle: Why We Find Ourselves Here

The Universe We Could Not Miss

Imagine a fish born in the ocean.

If that fish could think like a human being, it might one day say, "How amazing that the world is filled with water. Water surrounds me. Water supports me. Water enters my gills. Water lets me move, feed, survive, and reproduce. Surely this ocean was made for me."

The fish would be partly right.

The ocean does support the fish.

But the fish would also be making a mistake.

The fish exists there because that is where fish can exist. A fish cannot be born in dry desert sand and then wonder why the desert was not designed for fish. It can only find itself in an environment where fish life is possible.

Human beings often make a similar mistake with the universe.

We look around and say, "How amazing that the universe permits life. Gravity allows stars. Stars make heavy elements. Chemistry allows cells. Earth has liquid water. The laws of physics allow atoms, planets, and living beings. Surely all this was made for us."

But the first thing to understand is simple:

We could only observe a universe compatible with our existence.

If the universe did not permit observers, no one would be here to notice it.

This idea is called the anthropic principle.

It sounds simple, almost too simple. Yet it is one of the most important ideas in the discussion of fine-tuning, cosmic design, and God.

It does not answer every question.

It does not solve the deepest mysteries of cosmology.

It does not prove atheism by itself.

But it does weaken a common religious mistake: confusing the conditions that permit our existence with evidence that those conditions were created for us.

The universe may not be fitted to us.

We may be fitted to this universe.

The Observer's Trap

The anthropic principle begins with observer selection.

That phrase may sound technical, but the idea is simple.

Any observer who exists must observe conditions compatible with the existence of observers.

We cannot observe a universe where observers cannot exist.

We cannot be deeply surprised that our universe allows life, because if it did not allow life, there would be no one here to be surprised.

This does not mean the universe was designed.

It means our observations are filtered by the fact that we exist.

We do not observe reality from nowhere. We observe it from a particular place, at a particular time, under particular conditions that made our existence possible.

A person living on Earth should not be shocked that Earth has oxygen, water, and a suitable temperature range. If Earth did not have life-supporting conditions, that person would not be living here to observe them.

Similarly, human beings should not be shocked that the observable universe has conditions compatible with human existence. If the universe did not allow stars, atoms, chemistry, planets, or life, we would not be here.

The anthropic principle does not say the universe was made for us.

It says we can only find ourselves where we can exist.

This is a powerful correction to religious thinking.

Religion often says, “The universe fits us, therefore God designed it.”

Science replies, “We exist only where existence is possible.”

That is a much more careful statement.

Permission Is Not Purpose

One of the most important distinctions in this chapter is the difference between permission and purpose.

The universe permits life.

That does not mean the universe was created for life.

A mountain permits a tree to grow on one slope. That does not mean the mountain was made for that tree.

A cave permits bats to live inside it. That does not mean the cave was designed for bats.

A pond permits frogs to survive. That does not mean the pond exists for frogs.

Earth permits human life. That does not automatically mean Earth was created for human beings.

The universe permits life in at least one known place. That does not mean the universe was designed with human beings as its purpose.

Religious thinking often moves too quickly from permission to purpose.

It says, "Because life can exist, life must have been intended."

But this does not follow.

Many things happen because conditions allow them, not because they were planned. Snowflakes form when temperature, humidity, and molecular structure permit them. Crystals form when chemical and physical conditions allow them. Rivers form where water, gravity, and land shape flow. Stars form where gas clouds collapse under gravity.

No divine intention is needed.

Life may be similar.

Given suitable chemistry, energy, time, environment, and natural processes, life may arise and evolve. We do not yet know every detail of life's origin, and that belongs more fully to a later book in this series. But the point here is clear:

Possibility is not purpose.

The universe made life possible.

That does not prove it was made for life.

Why Our Existence Feels Like Evidence

Human beings naturally place themselves at the center of meaning.

This is not surprising. We experience reality from inside our own lives. Our pain feels central. Our love feels central. Our death feels enormous. Our families, fears, hopes, and memories fill our field of awareness.

So when we look at the universe, we often ask what it means for us.

Why are we here?

Why does the universe allow us?

Why do the laws permit our existence?

Why do the stars make the elements that form our bodies?

These questions feel personal.

Religion answers them personally.

God made you.

God planned you.

God created the universe so that life could appear.

God gave you purpose.

God placed you here.

This answer is emotionally powerful. It makes the universe feel like home. It makes existence feel intentional. It turns cosmic scale into personal meaning.

But emotional power is not evidence.

The anthropic principle reminds us that our perspective is biased by our existence. We are not neutral observers standing outside all possible realities. We are living beings who emerged in one physical reality and now wonder why that reality permits us.

The fact that we exist tells us something important:

The universe is compatible with our existence.

It does not automatically tell us:

The universe was designed for our existence.

That additional claim needs evidence.

Feeling chosen is not the same as being chosen.

The Puddle and the Hole

A familiar illustration may help.

Imagine a puddle of water in a hole in the ground. If the puddle could think, it might say, "This hole fits me perfectly. Every curve of the hole matches my shape. Surely the hole was made for me."

But the puddle has misunderstood the situation.

The puddle fits the hole because water takes the shape of the hole. The hole was not made for the puddle. The puddle adapted to the hole.

This is a useful image for the anthropic principle.

We may think the universe fits us. But in a deeper sense, we fit the universe because we arose within it. Our bodies, chemistry, biology, and minds were shaped by the conditions that made us possible.

The puddle analogy should not be pushed too far. Human life is far more complex than a puddle. Cosmology is far deeper than a hole in the ground. Fine-tuning is more serious than a simple shape-fitting example.

But the illustration captures the danger of religious thinking.

We look at conditions that permit us and assume those conditions were intended for us.

The puddle says, “The hole was made for me.”

Religion says, “The universe was made for us.”

Science-based atheism says, “We arose where arising was possible.”

That is less flattering.

But it is more reasonable.

The Weak Anthropic Principle

The weak anthropic principle is the most modest and most useful form of the idea.

It says, in simple language:

Our observations of the universe are conditioned by the fact that we exist as observers.

This means we should be careful when interpreting cosmic coincidences.

We live at a time in cosmic history when stars exist, planets exist, chemistry is possible, and biological life has had time to evolve. We should not be surprised that we do not live in an era before stars formed or after all stars have died. We could not exist in those periods.

We live in a region where conditions are compatible with life. We should not be surprised that we do not find ourselves floating in intergalactic emptiness or inside a star. We could not survive there.

We observe a universe with laws compatible with observers. We should not be surprised that we do not observe laws that make observers impossible.

This does not explain why the laws are what they are.

But it corrects our interpretation of what we observe.

It tells us that some features of our situation are not surprising evidence of design. They are necessary conditions for our existence as observers.

The weak anthropic principle is not mystical.

It is logical.

It says: do not treat your existence as independent from the conditions that made your existence possible.

That is simple.

And powerful.

Stronger Versions and Their Limits

There are also stronger versions of the anthropic principle.

Some versions suggest that the universe must have properties that allow observers to exist. Others speak in more speculative ways about the role of observers in the universe. Some versions move into philosophy, and some become controversial.

This stronger language should be handled carefully.

Used loosely, it can sound almost religious. It may seem to say that the universe had to produce life, or that observers are somehow necessary to the universe. Some people have used such language in ways that blur the line between science, philosophy, and metaphysics.

For this book, we do not need to defend any extravagant version of the anthropic principle.

The modest version is enough.

We only need the clear point that our observations are filtered by our existence.

We should not overstate the anthropic principle. It is not a complete explanation of the universe. It does not prove why the laws exist. It does not explain why there is something rather than nothing. It does not remove all fine-tuning questions.

But it does weaken a simple design claim.

It shows that the observation "the universe permits us" is not as surprising as it first appears, because only a universe permitting observers can be observed by observers.

This is not a trick.

It is a correction of perspective.

Science often advances by correcting perspective.

Earth feels still, but it moves.

The Sun appears to rise, but Earth rotates.

The sky looks like a dome, but it is space.

The universe feels made for us, but perhaps we arose because this universe permits us.

The Universe Is Not Especially Comfortable

The anthropic principle also helps us avoid exaggerating how life-friendly the universe really is.

We observe a universe compatible with our existence.

But that does not mean the universe is broadly friendly to us.

Most of it is not.

If human beings were placed almost anywhere in the universe without protection, they would die quickly. Empty space, intense radiation, extreme cold, crushing gravity, burning stars, toxic atmospheres, and airless worlds dominate cosmic reality.

Even Earth is not perfectly designed for comfort. It has earthquakes, volcanoes, floods, droughts, storms, disease, parasites, genetic defects, aging, pain, and death. Life survives, but it does not live in a perfectly arranged paradise.

This matters because religious fine-tuning arguments often speak as if the whole universe is lovingly prepared for human life.

The evidence does not show that.

The evidence shows that life exists in at least one small region where conditions allow it.

That is very different.

The anthropic principle gives a more modest interpretation.

We find ourselves in a life-compatible region because we could not exist in life-incompatible regions.

This does not mean life is unimportant. It means life is local, fragile, and dependent.

Science-based atheism can fully value life without pretending the whole universe was built for it.

In fact, life may become more precious when we understand how rare and fragile it may be.

Why the Anthropic Principle Does Not Solve Everything

The anthropic principle is useful, but it is not magic.

It does not answer every question.

It does not explain why there is a universe.

It does not explain why the laws of physics exist.

It does not explain why the constants have the values they do.

It does not prove a multiverse.

It does not fully explain fine-tuning.

It does not tell us whether life exists elsewhere.

It does not remove all mystery.

This must be admitted honestly.

Some atheists may be tempted to use the anthropic principle as if it solves everything. That would be a mistake. We should not replace religious overconfidence with atheist overconfidence.

The correct use of the anthropic principle is more careful.

It does not close the debate.

It changes the debate.

It says that our existence cannot be treated as a simple surprise because our existence is part of the conditions being discussed. It reminds us that observers can only appear where observers are possible. It removes some false astonishment from the design argument.

But it still leaves deep questions for physics and philosophy.

That is acceptable.

Science-based atheism does not need to answer every question immediately. It needs only to show that God is not required as the answer.

The anthropic principle helps do that.

It weakens the design claim without pretending to solve all cosmology.

The Anthropic Principle and the Multiverse

The anthropic principle becomes especially important if multiverse ideas are considered.

Suppose, for example, that there are many cosmic domains, perhaps with different constants, laws, or physical conditions. If most of them do not permit life, no observers would arise there. If a few do permit life, observers would naturally find themselves in those few life-permitting regions.

In that case, fine-tuning looks less surprising.

It would be like a lottery with many tickets. If only one ticket wins, the winner may feel amazed. But if a huge number of tickets exist, it is less surprising that one of them wins. The winner can still ask why this ticket won, but the fact that some ticket won is less mysterious.

This does not prove the multiverse.

The multiverse will be discussed more fully in the next chapter, and it must be treated with humility. It is not settled fact. It is debated. It depends on broader theories that are still being studied.

But the combination of anthropic reasoning and multiverse possibility shows that divine design is not the only possible response to fine-tuning.

Religion often says:

The universe permits life, therefore God designed it.

Science can reply: .

Perhaps observers naturally appear only in regions where observers can appear.

This reply may not answer everything.

But it is enough to block the claim that God is the only reasonable explanation.

Selection Effect Versus Divine Intention

The difference between selection effect and divine intention is central.

A selection effect occurs when the way something is observed affects what is observed.

For example, if a person asks why hospital patients are often sick, the answer is not that hospitals create all sickness. The answer is that sick people are more likely to be found in hospitals. The sample is selected by the condition being studied.

If a fisherman catches only large fish because his net has large holes, he should not conclude that the lake contains only large fish. His method selected what he could catch.

Similarly, we observe a universe compatible with observers because only such a universe can contain observers. Our existence selects the kind of universe we can observe.

Divine intention is different.

It says the universe has life-compatible features because a mind wanted life to exist.

That is a much larger claim.

The anthropic principle does not prove that divine intention is false. But it shows that we do not need divine intention to explain why observers observe observer-compatible conditions.

That is important.

The design argument often seems powerful because it treats our existence as an independent surprise. The anthropic principle shows that it is not independent. Our existence and our observations are connected.

We are not observers randomly dropped into any possible universe.

We are observers produced by this universe.

That changes everything.

We Are Not the Goal, We Are the Result

Religion often says that human beings are the goal of creation.

Science suggests something different.

We are not the goal.

We are the result.

We are the result of cosmic history, stellar evolution, planetary formation, chemistry, biology, mutation, natural selection, survival, extinction, and time. We are made from atoms forged in stars. We live on a planet shaped by physical processes. We think with brains built by evolution.

This does not make human life worthless.

It makes human life natural.

The universe did not need to have us in mind before we appeared. We came later. We emerged from conditions that permitted our existence. We are not the purpose of the universe, but we are one of the universe's most remarkable outcomes.

This is a hard shift for religious imagination.

Religion often begins with purpose and then explains the world from that purpose.

Science begins with reality and asks what processes produced what we see.

That is a more reliable method.

It may be less comforting at first. But it is more honest.

We do not need to be the planned goal of the cosmos in order to matter.

We matter because we are conscious, because we can suffer, because we can love, because we can understand, and because we can choose what kind of lives to build.

Meaning does not require cosmic design.

It requires conscious beings capable of meaning.

The Anthropic Principle and Human Humility

The anthropic principle is humbling.

It tells us that the universe may not be about us.

We are not the reason the universe has its laws. We are one result of those laws.

We are not outside reality, judging how well it was designed for us. We are inside reality, shaped by its conditions.

We are not the center.

We are consequences.

This can feel emotionally difficult.

Religion often tells human beings they are central, chosen, watched, loved, judged, and placed within a divine plan. The anthropic principle gives a cooler message: we find ourselves where beings like us can exist.

But that cooler message is not empty.

It is honest.

Human dignity does not require cosmic centrality. We do not need the universe to be made for us in order for our lives to matter. Love matters because conscious beings experience love. Suffering matters because conscious beings suffer. Truth matters because minds can seek it. Justice matters because lives can be harmed or protected.

Meaning does not require the universe to be arranged around us.

It requires beings capable of meaning.

We are such beings.

That is enough.

Science-based atheism asks us to grow beyond cosmic flattery. It asks us to accept that we are not the purpose of the universe, while still taking responsibility for the lives we live within it.

That is a mature worldview.

Does the Anthropic Principle Disprove God?

No.

The anthropic principle does not disprove God by itself.

It is important to be honest about this.

A believer can say that God created a universe in which observers would arise. A believer can say that the anthropic principle describes how we observe God's creation, not why creation exists. A believer can say that observer selection does not explain why there is a life-permitting universe in the first place.

These replies are possible.

But they do not defeat the main point.

The anthropic principle weakens the claim that life-permitting conditions are obvious evidence of design. It shows that our observation of such conditions is partly explained by the fact that we could not observe any other kind of universe.

God may still be added as belief.

But God is not forced by the evidence.

That distinction is central to this book.

Science-based atheism does not need to prove that every religious interpretation is logically impossible. It only needs to show that God is unnecessary as an explanation.

The anthropic principle helps show that.

It offers a natural correction to the feeling that the universe must have been designed for us.

The feeling may be powerful.

The inference is weak.

The Strongest Religious Reply

A thoughtful believer may say:

"The anthropic principle explains why we observe a life-permitting universe, but it does not explain why such a universe exists at all. It does not explain the laws of physics, the constants of nature, or why anything exists rather than nothing. Therefore, God remains the best explanation."

This is the strongest religious reply.

It is fair in one respect. The anthropic principle does not explain everything. It does not solve the ultimate question of existence. It does not provide a complete theory of laws and constants.

But the reply still moves too quickly to God.

The fact that one natural explanation is incomplete does not make a supernatural explanation true.

If the anthropic principle does not explain everything, the honest conclusion is that it does not explain everything. The honest conclusion is not "therefore God."

God still gives no mechanism.

God gives no testable model.

God gives no probability distribution.

God gives no explanation of why this universe and not another.

God gives no evidence that the designer is personal, loving, moral, or connected to religion.

The anthropic principle may be incomplete.

But God is not better explained.

God is simply asserted.

Evidence-Based Response

The evidence-based response is careful.

The anthropic principle does not answer every cosmic question. It does not fully explain fine-tuning, cosmic law, or ultimate existence. It should not be oversold.

But it does teach an important lesson.

We observe a universe compatible with our existence because only such a universe can contain observers. This means our existence is not independent evidence that the universe was designed for us. It is partly a selection effect.

This weakens simplistic design arguments.

It helps us see that life-compatible conditions are not automatically proof of divine intention. We may fit the universe because we arose within it, not because it was made for us.

God can still be added by belief, but God is not required by the anthropic principle.

Science-based atheism remains the more disciplined position:

Where evidence is incomplete, keep investigating.

Where explanation is partial, admit it.

Where mystery remains, do not turn it into God.

Chapter Summary

The anthropic principle begins with a simple idea: we can only observe a universe compatible with the existence of observers.

This does not solve every cosmic mystery. It does not explain why the laws of physics exist, why the constants have their values, or why there is something rather than nothing. But it does correct a common mistake in fine-tuning arguments.

We should not be surprised to find ourselves in conditions where we can exist. If those conditions did not hold, we would not be here to notice anything.

The universe may not be designed for us.

We may simply have arisen in the kind of universe where beings like us can arise.

This weakens the claim that life-permitting conditions are proof of God.

The anthropic principle is not a complete explanation, but it is a powerful correction of human-centered thinking.

Key Points

The anthropic principle says our observations are conditioned by our existence as observers.

We can only observe a universe compatible with observers.

Life-permitting does not mean designed for life.

Permission is not the same as purpose.

The weak anthropic principle is modest and useful.

Stronger forms should be treated carefully.

The anthropic principle does not solve every mystery.

It becomes especially important when multiverse ideas are considered.

Selection effect is not the same as divine intention.

We are not necessarily the goal of the universe. We may be one result of it.

God may be added as belief, but God is not required.

Transition to the Next Chapter

The anthropic principle shows why we should not be surprised to find ourselves in a universe compatible with observers.

But one question remains.

What if our observable universe is not all that exists?

What if there are many cosmic domains, perhaps with different conditions, different histories, or different values of physical constants?

This idea is called the multiverse.

It is fascinating. It is controversial. It is not proven. But it changes the fine-tuning debate in important ways.

In the next chapter, we will examine multiverse ideas and the importance of scientific humility.

Sources

Brandon Carter, writings on the anthropic principle

John D. Barrow and Frank J. Tipler, *The Anthropic Cosmological Principle*

Martin Rees, *Just Six Numbers*

Sean Carroll, *The Big Picture*

Stephen Hawking and Leonard Mlodinow, *The Grand Design*

Victor Stenger, *God: The Failed Hypothesis*

Stanford Encyclopedia of Philosophy, “Fine-Tuning”

Stanford Encyclopedia of Philosophy, “Cosmology and Theology”

Chapter 11: Multiverse Ideas and Scientific Humility

Perhaps Our Universe Is Not All There Is

For most of human history, the word universe meant everything.

All space.

All time.

All matter.

All stars.

All reality.

The universe was the total stage of existence. If someone asked where the universe came from, he was asking the largest possible question.

But modern cosmology has made even this word more difficult.

What if the universe we observe is not all that exists?

What if our visible cosmic region is only one part of a much larger reality?

What if there are other regions forever beyond our view?

What if different cosmic domains have different physical conditions?

What if our universe is not unique, but one member of a vast cosmic collection?

This idea is called the multiverse.

The word is dramatic. It can sound like science fiction. It can also be misused, exaggerated, or turned into a kind of atheist mythology. We must avoid that. The multiverse is not proven. It is not settled fact. It is not a simple replacement for God.

But it is important.

Why?

Because it changes the fine-tuning argument.

If there is only one universe, and it happens to permit life, some people may feel that design is the best explanation. But if there are many cosmic domains, perhaps with different conditions, then it becomes less surprising that at least one permits observers. We would naturally find ourselves in a life-permitting region, because we could not exist in any other kind.

This does not prove the multiverse.

But it shows something important:

God is not the only possible answer.

That is enough to weaken the religious argument.

Why the Multiverse Enters the God Debate

The multiverse enters the God debate because of fine-tuning.

As we saw in the previous chapters, the universe appears to have features that allow stars, atoms, chemistry, planets, and life. If certain constants or conditions were very different, life as we know it might not exist.

Believers often say:

The universe is fine-tuned.

Fine-tuning needs a fine-tuner.

Therefore, God designed the universe.

But this argument assumes that design is the best explanation for life-permitting conditions.

The multiverse offers another possibility.

If many cosmic domains exist, and if their conditions vary, then some may be lifeless, some may be unstable, some may never form stars, some may collapse quickly, and a few may allow complexity and life. Observers would appear only in those rare regions where observers can appear.

In that case, our presence in a life-permitting universe would not be shocking.

It would be expected.

This is similar to asking why we find ourselves on a planet with air, water, and suitable temperature. We find ourselves here because we could not survive on a planet without those conditions. We do not need to assume Earth was specially created for us. We can say that life arose where conditions allowed life.

Similarly, perhaps observers arise only in cosmic regions where physical conditions permit observers.

This does not answer every question.

It does not explain why there is a multiverse.

It does not explain why any laws exist.

It does not explain the deepest foundation of reality.

But it removes the religious claim that fine-tuning clearly points to one divine designer.

What "Multiverse" Means

The word multiverse does not always mean one thing.

Different scientists and philosophers use it in different ways.

Sometimes it means regions beyond our observable universe. The universe may extend far beyond what light has had time to reach us. If so, our observable universe is only the part we can see, not necessarily all that exists.

Sometimes it means many regions produced by cosmic inflation, where different regions may have different physical conditions.

Sometimes it means a landscape of possible universes suggested by certain interpretations of high-energy physics or string theory.

Sometimes it is connected to quantum mechanics, where different possible outcomes may be described in branching ways.

These ideas are not identical.

This matters because careless discussion creates confusion. A person may attack one version of the multiverse and think he has defeated all versions. Another person may defend one version and speak as if all versions are equally supported.

A serious book must be more careful.

For our purpose, we do not need to prove one specific multiverse theory. We only need to understand why multiverse ideas matter in the debate about God.

They matter because they show that life-permitting conditions do not automatically require design.

If many cosmic domains exist, observer selection becomes important. We would observe a life-permitting domain because only such a domain could contain observers.

That is the main point.

The multiverse is not proof of atheism.

It is a challenge to the claim that God is the only possible explanation for fine-tuning.

The Observable Universe and the Cosmic Horizon

To understand multiverse ideas, we must first understand the observable universe.

The observable universe is the region from which light has had enough time to reach us since the early universe became transparent. Because light travels at a finite speed, and because the universe has a finite age, there is a limit to what we can observe.

This limit is called a cosmic horizon.

Beyond that horizon, there may be more universe. We simply cannot see it directly, because light from those regions has not reached us and may never reach us.

This is not fantasy.

It follows from basic facts about light, time, and cosmic expansion.

Even if there is no exotic multiverse with different laws, it is already possible that reality extends far beyond the region we can observe. Our observable universe may be only one part of a larger whole.

This should make human beings humble.

Ancient religion was built from the small world people could see: land, sky, river, storm, Sun, Moon, visible stars, birth, disease, death, and dreams. Modern cosmology has shown that the visible world is only a small window into a much larger reality.

If even our observable universe may not be all there is, then religious claims based on ancient human experience become even less secure.

The universe is not limited to what human beings once imagined.

Reality may be larger than even our best current maps.

Inflation and Many Cosmic Regions

One major route to multiverse thinking comes from inflationary cosmology.

Inflation is the idea that the very early universe may have undergone an extremely rapid expansion. This idea was developed to address certain puzzles in cosmology, such as why the universe looks so uniform on large scales and why its geometry appears so close to flat.

Inflation has been scientifically influential, though not every detail is settled. Some versions of inflation suggest that inflation may not stop everywhere at once. It may end in some regions while continuing in others. This can lead to the idea of many bubble-like cosmic regions, each with its own history.

This picture is sometimes called eternal inflation.

The language can sound strange, but the basic idea is simple: what we call our universe may be one region within a much larger inflating reality.

If different regions have different physical conditions, then fine-tuning may be partly explained by selection. Most regions may not support life. We are in a region that does because only such a region could produce beings like us.

Again, this is not proven.

But it is scientifically motivated in a way that "God did it" is not.

Inflation arises from attempts to solve real problems in cosmology. It connects to models, observations, mathematics, and physical theory. It may be right, wrong, or incomplete. But it belongs to scientific investigation.

God does not enter the calculation.

That difference matters.

The multiverse may be uncertain science.

God is not science at all.

The String Landscape

Another multiverse-related idea comes from string theory.

String theory is an ambitious attempt to describe fundamental physics in a deeper way. It suggests that the basic ingredients of reality may not be point-like particles, but tiny vibrating strings or other extended objects. The theory is mathematically rich, but difficult to test directly.

One idea that emerged from string theory is sometimes called the landscape.

This refers to a huge number of possible solutions or vacuum states, each potentially corresponding to different physical constants or laws. If such a landscape is real, then different cosmic regions might realize different possibilities.

This could help explain why our universe has the values it does.

We observe these values not because a divine mind chose them, but because we exist in a region where those values permit observers.

This is a powerful idea.

But it must be handled with caution.

String theory has not been experimentally confirmed in the way general relativity or Big Bang cosmology has. The landscape idea is debated. Some physicists see it as promising. Others criticize it as difficult to test or too flexible.

A fair book must admit this.

We should not pretend that the string landscape has solved fine-tuning.

It has not.

But it still matters because it shows that fine-tuning may have natural explanations beyond our current knowledge.

The design argument says: we do not know why the constants are life-permitting, therefore God.

Science says: there may be deeper physical reasons, and we should keep looking.

That difference is the heart of the matter.

Quantum Many Worlds

Another idea sometimes included under the multiverse label comes from quantum mechanics.

In one interpretation, often called the many-worlds interpretation, different possible outcomes of quantum events are all realized in branching realities. This is not the same as inflationary multiverse or string landscape. It comes from a different part of physics and raises different questions.

This idea is fascinating, but it is not necessary for the main argument of this book.

It is mentioned here only to make one point clear: the word multiverse covers several different ideas, not one single theory.

A reader should be cautious when hearing arguments for or against "the multiverse." Which multiverse? Inflationary? String landscape? Quantum many-worlds? Regions beyond the observable universe? These are not the same thing.

For the fine-tuning debate, the most relevant versions are those in which many cosmic regions may exist with different physical conditions.

The many-worlds interpretation may or may not be correct. It may or may not have anything important to say about fine-tuning. But it shows again that modern physics is not bound by the small, human-shaped categories of ancient religion.

Reality may be stranger than our inherited stories.

That does not prove every speculative idea true.

It only reminds us to remain intellectually humble.

Why the Multiverse Is Not Proven

The multiverse is fascinating, but it is not proven.

This must be said plainly.

We cannot directly observe other universes or distant cosmic domains beyond our horizon. We cannot send instruments outside our observable universe. We cannot compare our universe with a known catalog of other universes. We do not yet have direct evidence that different regions with different constants actually exist.

This is why some critics say the multiverse is speculative.

They are not entirely wrong.

A scientific idea can be serious and still unproven. It can be mathematically motivated and still uncertain. It can arise from real physics and still lack direct confirmation.

The multiverse should not be presented as a fact in the same way that the cosmic microwave background is a fact.

The cosmic microwave background is observed.

The expansion of the universe is observed.

The multiverse is a possible implication of certain theories.

This distinction matters.

Science-based atheism must remain honest. It should not turn unproven ideas into dogma merely because they are useful against religion. That would be intellectually dishonest.

If we criticize religious believers for turning mystery into certainty, we must not do the same thing.

The right position is this:

The multiverse is not proven, but it is a serious possibility in some scientific contexts, and it shows that fine-tuning does not force us to accept God.

That is careful.

That is enough.

Is the Multiverse Just Another Faith?

Religious critics often say that belief in the multiverse is just another faith.

They argue that if atheists can believe in unseen universes, then believers can believe in an unseen God. They say the multiverse is no more scientific than theology.

This objection has some force, but only partly.

The multiverse should not be accepted by blind faith. It should not be treated as proven. It should not be used as a slogan.

But it is not exactly the same as God.

Multiverse ideas usually arise from mathematical models and physical theories developed for reasons independent of religion. Inflation, quantum theory, and high-energy physics generate questions that may lead to multiverse possibilities. These ideas are connected to scientific frameworks that also make contact with observation in other ways.

God is different.

God is usually introduced as a personal, conscious, intentional, supernatural being beyond space and time.

God is not derived from a tested physical model. God gives no measurable predictions. God does not tell us what constants should exist. God does not explain why this universe rather than another. God does not produce a research program comparable to cosmology.

So the multiverse and God are not equivalent.

However, the critic is right about one thing: the multiverse should be handled with humility.

It is a possibility, not a certainty.

It may help explain fine-tuning, but it has not been fully established.

Science-based atheism should not depend entirely on the multiverse. The case against God does not collapse if the multiverse is wrong. Fine-tuning still does not prove God, because probability problems, deeper law, selection effects, and the problem of the designer remain.

The multiverse is one possible natural path.

It is not the only one.

Why God Is Not a Better Explanation

Some believers say the multiverse is speculative, so God is a better explanation.

But this does not follow.

If one explanation is uncertain, it does not mean another explanation is true.

The multiverse may be speculative.

But God is more speculative.

The multiverse is at least connected to physical theory. God is not. The multiverse, in some versions, may have indirect consequences or theoretical connections. God has no comparable scientific structure. The multiverse does not require a supernatural mind beyond space and time. God does.

Also, God explains fine-tuning poorly.

If God designed the universe for life, why is most of the universe hostile to life?

If God fine-tuned reality for human beings, why did the universe exist for billions of years before human beings appeared?

If God wanted life, why use a universe filled with radiation, extinction, cosmic violence, and vast emptiness?

If God chose the constants, why these constants and not better ones?

If God is the designer, who explains the designer?

The multiverse has problems.

But God has larger problems.

God is not a better explanation merely because another idea is uncertain.

A weak answer does not become strong because a different answer is incomplete.

The honest position remains: we do not yet know the full answer to fine-tuning.

But we have no good reason to call the answer God.

The False Choice: Multiverse or God

The debate is often framed badly.

Some believers say, "Either God designed the universe, or atheists must believe in the multiverse."

This is a false choice.

The alternatives are broader.

Fine-tuning might be explained by deeper physical law.

It might be explained by relationships among constants we do not yet understand.

It might be explained partly by observer selection.

It might be explained by a multiverse.

It might be explained by future physics.

It might involve brute facts about reality.

Or the final answer may be something no one has yet imagined.

The point is not that we already know the answer.

The point is that God is not forced.

Science-based atheism does not require dogmatic belief in the multiverse. It requires only disciplined resistance to unsupported divine claims.

A mystery does not become God just because one possible natural explanation is uncertain.

This is important.

The reader should not leave this chapter thinking, "The multiverse proves atheism."

That would be too strong.

The reader should leave thinking, "Fine-tuning does not prove God, because there are serious natural possibilities, and the honest answer is still open."

That is the better conclusion.

The Danger of Atheist Overconfidence

A book defending science-based atheism must still be careful.

Atheists sometimes overstate the multiverse. They may speak as if it has already solved fine-tuning completely. They may treat it as obvious fact. They may use it as a quick escape from difficult questions.

That is a mistake.

The goal of science-based atheism is not to win arguments by exaggeration.

The goal is to understand reality honestly.

If the multiverse is uncertain, say it is uncertain.

If fine-tuning remains a real question, say it remains a real question.

If physics is incomplete, admit it.

There is no need to pretend that every mystery has vanished.

Atheism does not require final knowledge.

It requires refusal to insert God without evidence.

This is a mature position.

Religion often says: we do not know, therefore God.

Atheism should not say: we do not know, therefore multiverse.

The better position is: we do not know fully, but natural explanations remain more disciplined, more fruitful, and more consistent with the history of science than supernatural explanations.

Scientific humility is not a weakness.

It is one of the strongest reasons to trust science.

Science can say, “We may be wrong.”

Religion often says, “We cannot be wrong because God revealed it.”

That difference matters.

Scientific Humility and Religious Certainty

The multiverse debate reveals a deeper contrast between science and religion.

Science lives with uncertainty.

It proposes models.

It tests.

It revises.

It doubts.

It argues.

It changes.

It follows evidence where evidence leads.

Religion often begins with certainty. God exists. God created. God intended. God designed. God revealed. God commands. God judges.

Then it tries to fit the world into that certainty.

This is why scientific humility is intellectually powerful.

When scientists discuss the multiverse, they debate it. Some defend it. Some criticize it. Some think it is promising. Some think it is too speculative. This open disagreement is not a failure. It is how science protects itself from dogma.

Religious certainty often does the opposite. It protects belief from correction.

In the debate over fine-tuning, the believer often says, "God is the answer." Science says, "There are several possible paths, and we do not yet know which is right."

Which response is more honest?

The second.

It may be less emotionally satisfying. It may not give the comfort of final certainty. But it is more faithful to reality.

Science-based atheism is not the claim that we know everything.

It is the claim that we should not pretend to know what we do not know by naming the unknown God.

The Multiverse and Human Importance

Multiverse ideas are also humbling.

Religion often makes human beings central to cosmic purpose. God creates the universe. God creates humans. God watches humans. God judges humans. God saves humans. The entire structure of reality becomes tied to human destiny.

The multiverse, if true, would make that picture even less plausible.

Our observable universe would not be the whole of reality. It would be one region in a larger structure. Our galaxy would be one among billions. Our planet would be one among countless planets. Our species would be one late, fragile form of life. Our religious arguments would be local products of one small world.

This does not make human life worthless.

It makes human arrogance harder to defend.

A multiverse would not remove meaning from human life. Meaning does not require cosmic centrality. Love, suffering, truth, justice, and compassion matter because conscious beings experience them.

But the idea that all reality exists mainly for human religious drama becomes increasingly difficult.

Science repeatedly teaches the same lesson.

We are not central.

We are not cosmic rulers.

We are not the purpose of all stars.

We are not the obvious goal of all existence.

We are part of nature.

This is not humiliation.

It is maturity.

The Unknown Is Not Owned by Religion

The multiverse remains uncertain.

Fine-tuning remains debated.

Cosmic origins remain difficult.

Physics remains incomplete.

But the unknown is not owned by religion.

This sentence should guide the whole book.

For centuries, religion has tried to occupy the unknown. If people did not know what caused thunder, God was placed there. If they did not know what caused disease, demons or divine punishment were placed there. If they did not know how species formed, God was placed there.

If they did not know how the universe began, God was placed there.

But history shows that this is a dangerous place for God.

The unknown changes.

Science advances.

Gaps close.

Mysteries become problems.

Problems become research.

Research becomes explanation.

The multiverse may or may not be part of the final answer. But its very possibility reminds us that the unknown is open. It is not automatically religious territory.

Science does not need to have every answer today to reject unsupported divine certainty.

It only needs to say:

There are natural possibilities.

There is no good evidence for God.

Therefore, God is not required.

That is enough.

The Strongest Religious Reply

A thoughtful believer may say:

"The multiverse is unproven. No one has observed other universes. It cannot be directly tested. It is used mainly to avoid the obvious conclusion that our universe was designed. Belief in a multiverse may require as much faith as belief in God. Therefore, fine-tuning still points more strongly to a creator."

This is a serious reply.

It correctly warns against overclaiming. It is true that the multiverse is not directly observed. It is true that some versions are difficult to test. It is true that atheists should not use it carelessly.

But the reply still does not prove God.

First, the multiverse is not the only alternative to design.

Second, fine-tuning probabilities remain uncertain.

Third, deeper laws may explain what now appears finely tuned.

Fourth, observer selection effects matter.

Fifth, God is not a simple explanation.

Sixth, God gives no testable mechanism.

Seventh, a designer of universes would itself require explanation.

Eighth, even if design were suspected, it would not prove any specific religion.

The multiverse does not need to be proven in order to weaken the claim that God is the only possible explanation.

It is enough that design is not forced.

Evidence-Based Response

The evidence-based response is careful.

Multiverse ideas are not proven. They should not be presented as settled science. They should be discussed with humility.

But they are scientifically relevant because some arise from serious attempts to understand inflation, quantum theory, and fundamental physics. They show that fine-tuning may have natural explanations beyond simple chance or divine design.

This does not mean the multiverse is true.

It means God is not required.

The religious argument from fine-tuning depends on the claim that design is the best or only explanation. Multiverse ideas, deeper laws, probability problems, and anthropic reasoning all weaken that claim.

Science-based atheism does not need to replace God with dogmatic multiverse belief. It only needs to keep the question open and refuse unsupported supernatural certainty.

The honest answer is:

We do not yet know the full explanation of fine-tuning.

But we have no good reason to conclude that the answer is God.

Chapter Summary

The multiverse is the idea that our observable universe may not be all that exists. There may be many cosmic regions or domains, perhaps with different conditions, laws, or constants.

This idea matters because of fine-tuning. If many cosmic domains exist, then it becomes less surprising that at least one permits observers. We would naturally find ourselves in a life-permitting region because only such a region could contain beings like us.

The multiverse is not proven and should not be treated as settled fact. It must be discussed with humility. But it is not the same as religious faith. Some multiverse ideas arise from scientific theories such as inflationary cosmology, quantum theory, and high-energy physics.

The multiverse does not prove atheism.

But it weakens the claim that fine-tuning proves God.

God is not required.

Key Points

The multiverse means that our observable universe may not be all that exists.

Different multiverse ideas come from different areas of physics and cosmology.

The observable universe is limited by a cosmic horizon.

Inflationary cosmology may suggest many cosmic regions.

The string landscape may suggest many possible physical conditions.

The multiverse is not proven and should not be overstated.

It is not equivalent to belief in God because it arises from scientific models, not revelation.

God is not a better explanation merely because the multiverse is uncertain.

Fine-tuning does not force a design conclusion.

Scientific humility is stronger than religious certainty.

The unknown is not owned by religion.

Transition to the Next Chapter

The multiverse asks us to imagine reality larger than our observable universe.

But there is another reason to be cautious about simple creator arguments.

At the deepest levels of physics, common sense often fails.

Words such as cause, before, nothing, beginning, and existence may not behave in the way ordinary human life suggests. The early universe may not fit the simple mental pictures used in many religious arguments.

In the next chapter, we will enter the strange world of quantum cosmology and the limits of common sense.

Sources

Alan Guth, *The Inflationary Universe*

Andrei Linde, writings on inflationary cosmology and eternal inflation

Brian Greene, *The Hidden Reality*

Sean Carroll, *The Big Picture*

Martin Rees, *Just Six Numbers*

Leonard Susskind, *The Cosmic Landscape*

Stephen Hawking and Leonard Mlodinow, *The Grand Design*

Stanford Encyclopedia of Philosophy, “Fine-Tuning”

Chapter 12: Quantum Cosmology and the Failure of Common Sense

When Ordinary Words Begin to Break

Human beings live by ordinary words.

Before.

After.

Cause.

Effect.

Something.

Nothing.

Beginning.

End.

These words guide daily life. A child is born after pregnancy. A fire starts after a match is struck. A cup falls because a hand knocks it from the table. A wound bleeds because skin and blood vessels are injured. A house exists because people built it. A meal exists because someone cooked it.

In ordinary life, cause comes before effect.

A beginning seems to need something before it.

Something does not seem to come from nothing.

This is common sense.

And common sense is useful.

But the early universe is not ordinary life.

At the deepest edge of cosmology, ordinary words begin to tremble. Time may not behave the way we imagine. Cause may not mean what it means in daily experience. Nothing may not be the empty absence we picture in our minds. Beginning may not mean a moment inside a larger clock.

This is where quantum cosmology enters.

Quantum cosmology asks what happens when the universe itself is studied using the ideas of quantum physics. It tries to understand the earliest possible conditions of the universe, where general relativity and quantum mechanics must somehow meet.

This is difficult science.

It is not complete.

It is not settled.

No honest writer should pretend otherwise.

But it teaches one powerful lesson:

The universe is not required to obey human common sense.

This matters greatly for religion.

Many arguments for God depend on everyday intuition. Everything must have a cause. Nothing can come from nothing. Time must have a first moment. A beginning must have a beginner. A law must have a lawgiver.

These statements sound strong because they feel natural.

But feeling natural is not the same as being true.

At the edge of the universe, common sense may fail.

And when common sense fails, God should not be used to fill the confusion.

The Limits of Human Intuition

Human intuition evolved for survival, not cosmology.

Our minds are good at handling middle-sized objects in ordinary environments. We understand stones, trees, animals, water, fire, food, danger, faces, tools, roads, and bodies. We understand practical cause and effect because it helps us survive.

If smoke rises, there may be fire.

If an animal growls, there may be danger.

If a child has fever, there may be illness.

If food smells rotten, it may be unsafe.

This kind of intuition is useful.

But the universe is not limited to the scale of human survival.

We did not evolve to understand the speed of light.

We did not evolve to understand curved space-time.

We did not evolve to understand quantum uncertainty.

We did not evolve to understand black holes.

We did not evolve to understand the earliest universe.

Again and again, science has shown that intuition can mislead us.

Earth feels still, but it moves.

The Sun appears to travel around Earth, but Earth rotates.

Solid matter feels completely solid, but at the atomic level it is mostly structured space.

Time feels absolute, but relativity showed that time is more subtle.

Particles feel as if they should behave like tiny balls, but quantum physics shows a stranger reality.

This should make us cautious.

Religious arguments often take ordinary human intuition and stretch it to the whole universe. They say the universe must have a maker because things in ordinary life have makers. They say the universe must have a cause because events in ordinary life have causes.

But the universe is not an ordinary object.

Cosmic origin is not carpentry.

The beginning of space-time is not like building a chair.

Quantum Physics in Plain Language

Quantum physics studies nature at very small scales.

At those scales, matter and energy do not behave in the simple way ordinary objects do. Particles can behave in wave-like ways. Events can be described by probabilities. Measurement matters. Uncertainty is built into the theory in ways that are not merely due to human ignorance.

Quantum physics is not common sense.

But it works.

It has been tested with extraordinary success. It underlies modern electronics, lasers, semiconductors, medical imaging, and much of modern technology. It may feel strange, but its strangeness does not make it false.

This is an important lesson.

Reality can be strange and still be real.

Religion often explains mystery by making it personal. If something is hard to understand, perhaps a mind is behind it. If the universe is strange, perhaps God made it. If existence is mysterious, perhaps divine will explains it.

Science takes a different path.

It studies the strangeness.

Quantum physics did not say, "This is too strange, therefore God."

It asked what the evidence showed.

That attitude should guide cosmology also.

The early universe may be deeply strange. But strangeness is not evidence for God. It is evidence that reality is deeper than ordinary imagination.

Where Relativity and Quantum Theory Meet

Modern physics rests on two great pillars.

General relativity explains gravity, space-time, and the large-scale universe. It helps us understand planets, stars, black holes, gravitational waves, cosmic expansion, and the shape of the cosmos.

Quantum mechanics explains the small-scale behavior of matter and energy. It helps us understand atoms, particles, radiation, fields, chemistry, electronics, and much of modern technology.

Both theories are enormously successful.

But they do not yet fit together completely.

This creates a problem at the earliest universe.

If we trace the history of the universe backward, we reach conditions of extreme density, extreme temperature, and tiny scale. In that region, gravity matters intensely, but quantum effects also matter. General relativity alone is not enough. Quantum mechanics alone is not enough.

A deeper theory of quantum gravity is needed.

We do not yet have a complete and confirmed theory of quantum gravity.

This is not a failure of science.

It is the honest frontier of science.

The religious temptation is to look at this frontier and say, "There is God."

But that is the old mistake.

A frontier is not an altar.

A gap in present physics is not evidence for a supernatural mind.

It is a sign that more work is needed.

What Quantum Cosmology Tries to Do

Quantum cosmology tries to apply quantum ideas to the universe as a whole, especially to its earliest state.

This is one of the hardest areas of modern thought.

Some models suggest that the universe may not have begun in the simple way ordinary language imagines. Some suggest that time may behave differently near the earliest boundary. Some propose that the universe could arise from a quantum state. Some propose that the universe may not have a sharp beginning in time. Some propose cyclic or bouncing cosmologies. Some remain highly speculative.

The details are complex and debated.

But one lesson is clear enough for this book:

The old religious picture of God standing outside time and starting the universe like a craftsman starting a machine is not required by modern physics.

It may be emotionally familiar.

It may be theologically comforting.

But it is not forced by science.

Quantum cosmology makes simple creator arguments less secure, not more secure.

It does not give us permission to say, "Science has solved everything."

It gives us reason to say, "The beginning may not be what common sense imagines."

That is enough to weaken the simple God argument.

The Problem With "Before the Universe"

People often ask, "What happened before the universe began?"

This question feels natural.

But it may not be meaningful in the ordinary way.

If time is part of the universe, then asking what happened before the universe may be like asking what is north of the North Pole. The words are familiar, but the question may not fit the reality.

This does not mean the question is stupid.

It means the question may need to be reformulated.

In ordinary life, a beginning occurs inside time.

A song begins.

A fever begins.

A journey begins.

A surgery begins.

But the beginning of the universe, if that phrase is even fully correct, may not be a beginning inside a larger time. It may involve the origin, boundary, or transformation of time itself.

This creates a serious problem for religious arguments.

They often imagine God existing first, then creating the universe.

But first is a time word.

If time itself is part of the universe, what does it mean for God to exist before time?

Theologians may answer that God is timeless. But then new questions arise. How does a timeless being act? How does a timeless being decide? How does a timeless will produce a temporal universe? Does intention require sequence? Does creation involve change?

These are not simple questions.

Religion often presents God as a simple answer.

But at this level, God becomes another mystery.

Physics "Nothing" and Philosophical Nothing

The word nothing causes enormous confusion.

In ordinary speech, nothing may mean empty space. A room with no furniture is called empty. A box with no objects is said to contain nothing. A region of space without stars or planets may be called empty space.

But empty space is not absolute nothing.

In physics, what seems empty can still have structure. It can contain fields. It can have energy. It can obey laws. It can have quantum activity. A vacuum in physics is not the same as philosophical nothingness.

Philosophical nothingness means no space, no time, no matter, no energy, no laws, no fields, no possibilities, no anything.

This distinction is crucial.

When physicists speak about a universe arising from "nothing," critics often object that this is not really nothing. Sometimes the critics are correct. In many scientific discussions, the "nothing" being described is not absolute nothingness. It is a physical state, a quantum vacuum, or a law-governed framework.

Science must be careful with language.

But religion must be careful too.

When believers say God created the universe from nothing, they do not really begin with absolute nothing either. They begin with God. And God is not nothing. God is an infinitely powerful, timeless, spaceless, immaterial mind with creative ability.

That is not nothing.

That is a very large something.

So the religious claim does not solve the problem of nothingness. It replaces nothing with God, then exempts God from explanation.

Science may not yet fully explain why there is something rather than nothing.

But God does not clearly explain it either.

Does Quantum Physics Mean Something Comes From Nothing?

Some people say quantum physics shows that something can come from nothing.

Others say this is misleading.

Both sides can be careless.

Quantum physics does show that at very small scales, reality behaves in ways that surprise common sense. It also shows that vacuum states are not simple empty nothingness. Quantum fluctuations and fields belong to a physical framework. They are not absolute nothing.

So we should not say too quickly that physics has fully explained how absolute nothing becomes something.

That would be an overclaim.

But we should also not let religion misuse the difficulty.

The fact that physics has not fully solved the question of ultimate origin does not prove God.

A mystery in physics is not evidence for theology.

The honest position is this:

Quantum physics makes ordinary ideas of cause, time, and nothing more complicated than religious arguments often admit. It does not give us a simple final answer to existence. But it weakens the claim that common sense proves a creator.

This is enough for the purpose of this chapter.

Science-based atheism does not need to pretend that physics has solved everything.

It only needs to show that God is not required.

The Universe May Not Need a Temporal Cause

In ordinary life, causes happen in time.

A cause comes before an effect.

But if time itself is part of the universe, the universe may not need a cause in the ordinary temporal sense.

This is difficult because the human mind wants a story.

First there was God.

Then God created.

Then the universe existed.

That is a story. It has sequence. It has a person. It has intention. It feels understandable.

But reality may not work like a story.

Modern physics suggests that time is not a universal background independent of everything else. Time is related to space, motion, gravity, and perhaps quantum conditions at the deepest level. If so, then the origin of the universe may not be explainable by a simple before-and-after narrative.

A non-temporal explanation, if one exists, may be physical rather than divine.

Perhaps the universe is self-contained in some way.

Perhaps the earliest state is not a point requiring an external trigger.

Perhaps deeper laws or quantum gravity will change the question.

Perhaps our language is inadequate.

We do not yet know.

But not knowing does not prove God.

The religious story is easier to picture because it is human-shaped.

A cosmic parent acts.

A divine mind chooses.

A maker makes.

But the universe does not need to be human-shaped.

That is one of the great lessons of science.

The No-Boundary Idea

Stephen Hawking and James Hartle proposed an idea known as the no-boundary proposal.

The details are technical, and this book does not need to explore the mathematics. But the general idea is important.

The proposal suggests that the universe may be finite but without a boundary in time in the ordinary sense. Hawking compared it loosely to the surface of Earth. Earth's surface is finite in area, but if you travel north, you do not eventually reach a sharp edge where Earth's surface suddenly begins. The North Pole is not a boundary beyond which the surface ends. It is a point where the direction north no longer continues.

Similarly, in the no-boundary idea, asking what happened before the universe may be like asking what lies north of the North Pole.

This does not prove there is no God.

It is a scientific proposal, not a final metaphysical conclusion.

But it is important because it shows that the universe may not need the kind of beginning religious imagination assumes. The earliest universe may not be a sharp edge requiring God to push it into existence.

The no-boundary proposal is debated. It should not be exaggerated. But it is a serious example of how physics can challenge the simple creator picture.

Religion says there must be a beginning that needs a beginner.

Physics asks whether beginning itself has been misunderstood.

That is a powerful difference.

Other Natural Possibilities

The no-boundary idea is not the only possible approach.

Some models suggest a bouncing universe, in which a previous phase may have led into the expansion we observe. Some cyclic models imagine repeated cosmic phases. Some quantum gravity approaches suggest that the classical Big Bang singularity may be replaced by a different physical description.

These ideas are not settled.

Some may be wrong.

Some may be incomplete.

Some may never become the final answer.

But their existence matters for the philosophical debate.

They show that the religious picture is not the only possible way to think about cosmic origin. The choice is not between "God created the universe" and "nothing makes sense." There are serious natural possibilities under investigation.

A mature science-based atheism does not need to declare any one speculative model final.

It simply says that difficult physics should remain open to investigation.

God should not be inserted as a shortcut.

Common Sense and the First-Cause Argument

The first-cause argument depends heavily on common sense.

Everything that begins has a cause.

The universe began.

Therefore, the universe has a cause.

That cause is God.

But quantum cosmology makes us question each step.

What does it mean for the universe to begin?

Does time itself begin?

Can causation apply without time?

Is the universe an object inside a larger reality, or the total physical system?

Does quantum physics allow kinds of explanation that do not resemble ordinary cause and effect?

Can a universe be self-contained?

Can the earliest state be described without a prior temporal cause?

These questions do not prove atheism by themselves.

But they show that the religious argument is not as simple as it appears.

The first-cause argument often has the tone of certainty. It sounds like an obvious truth. But its certainty comes from ordinary intuition, not from demonstrated knowledge of the earliest universe.

Science-based atheism does not need to give a final theory here.

It only needs to expose the weakness of the religious leap.

If common sense is unreliable at cosmic and quantum scales, then common-sense arguments for God are unreliable too.

Why God Should Not Be Inserted Into Difficult Physics

Difficult physics creates temptation.

When a subject becomes hard, people may want a simple answer.

God is the simplest sounding answer of all.

What caused the universe?

God.

Why are there laws?

God.

Why is there something?

God.

Why does anything exist?

God.

But simplicity of language is not the same as explanatory power.

God is a short word, but it hides enormous problems.

How does God exist?

How does God create?

How does God act without time?

How does God create matter without matter?

How does God produce physical law?

Why does God create this universe?

Why does God remain hidden?

Why does the universe look as if it operates naturally?

Difficult physics should not be solved by difficult theology disguised as simplicity.

If science does not yet know something, the correct answer is not "God." The correct answer is "science does not yet know."

This may feel less satisfying, but it is more honest.

History repeatedly shows that God placed in the difficult places of knowledge later retreats. The safe habit is not to insert God into the unknown.

The safe habit is to keep asking.

What This Chapter Does Not Claim

This chapter should not be misunderstood.

It does not claim that quantum cosmology has solved the origin of the universe.

It does not claim that the no-boundary proposal is proven.

It does not claim that bouncing or cyclic models are established fact.

It does not claim that physics has explained absolute nothingness.

It does not claim that all philosophical questions are finished.

That would be dishonest.

The correct claim is more careful and stronger:

Quantum cosmology shows that the earliest universe may not fit ordinary human ideas of cause, time, beginning, and nothing. Therefore, simple religious arguments based on common sense are not reliable.

This is enough.

Science-based atheism does not need false certainty.

It needs disciplined honesty.

Where science knows, it explains.

Where science does not yet know, it keeps the question open.

What it must not do is surrender the unknown to God.

The Religious Reply

A thoughtful believer may say:

"Quantum cosmology is speculative. It does not explain why there are laws of physics. It does not explain why there is something rather than nothing. It does not disprove God. The universe still requires an ultimate explanation, and God remains the best answer."

This reply has some strength.

It is true that quantum cosmology is incomplete. It is true that we do not yet have a final theory of quantum gravity. It is true that physics has not fully explained why there is something rather than nothing. It is true that quantum cosmology does not prove that no God exists.

But the reply still does not prove God.

It only points to remaining mystery.

A remaining mystery is not evidence for a divine mind.

If God is offered as the ultimate explanation, God must also be explained or justified. Calling God timeless, necessary, or beyond understanding does not make God a scientific explanation. It gives no mechanism, no testable prediction, no measurable evidence, and no clear connection to the God of religion.

Quantum cosmology may be incomplete.

But theology is not complete science.

The honest answer is not God.

The honest answer is that the deepest questions remain open, and natural inquiry remains the best method we have.

Evidence-Based Response

The evidence-based response is careful.

Quantum cosmology does not yet give a final answer to the origin of the universe. It is an active, difficult, and incomplete field. We should not exaggerate it.

But quantum cosmology does show that ordinary ideas of cause, time, beginning, and nothing may not apply simply to the earliest universe. This weakens religious arguments that depend on common sense.

The universe may not require a cause in the same way that a house requires a builder.

The beginning of time may not be like the beginning of a story.

Physics nothing is not the same as philosophical absolute nothing.

A boundary in our current knowledge is not proof of God.

Science-based atheism does not claim that every mystery is solved. It claims that mystery should not be turned into God without evidence.

That claim remains strong.

Chapter Summary

Quantum cosmology examines the earliest universe using ideas from quantum physics and cosmology. It is a difficult and incomplete field, but it has major philosophical importance.

It shows that ordinary ideas such as before, cause, beginning, and nothing may not apply simply to the universe as a whole. Human common sense, useful in daily life, often fails at cosmic and quantum scales.

This weakens simple creator arguments. The universe may not need a cause in the same way ordinary objects need causes. The beginning of time may not be like the beginning of an event inside time. Physics "nothing" is not the same as absolute philosophical nothingness.

Quantum cosmology does not disprove God completely.

But it does show that God is not required by the failure of common sense.

Mystery remains.

God is not the answer.

Key Points

Human common sense evolved for ordinary life, not cosmic origins.

Quantum physics shows that reality can behave in ways that defy intuition.

Quantum cosmology remains incomplete and should not be overclaimed.

Cause, before, beginning, and nothing may not apply simply to the early universe.

Physics "nothing" is different from philosophical absolute nothing.

The no-boundary proposal challenges simple ideas of a first moment.

Other natural models, such as bouncing or cyclic cosmologies, show that the creator story is not the only option.

Religious first-cause arguments rely heavily on ordinary intuition.

Difficult physics should not be filled with God.

Science-based atheism accepts uncertainty without turning it into theology.

Transition to the Next Chapter

Quantum cosmology shows that the earliest universe may not fit ordinary ideas of cause, time, and beginning.

This brings us to one of the most famous scientists of modern times: Stephen Hawking.

Hawking spent much of his life studying black holes, singularities, time, and the origin of the universe. He did not answer every question. He did not end the God debate. But he helped millions understand that the universe may be explainable through physical law without divine creation.

In the next chapter, we will examine Stephen Hawking and a universe from law.

Sources

Stephen Hawking, *A Brief History of Time*

Stephen Hawking and Leonard Mlodinow, *The Grand Design*

James Hartle and Stephen Hawking, writings on the no-boundary proposal

Sean Carroll, *From Eternity to Here*

Brian Greene, *The Fabric of the Cosmos*

Alan Guth, *The Inflationary Universe*

Kip Thorne, *Black Holes and Time Warps*

Stanford Encyclopedia of Philosophy, “Cosmology and Theology”

Chapter 13: Stephen Hawking and a Universe from Law

The Mind That Traveled Beyond the Body

The image became unforgettable.

A thin man in a wheelchair. A body almost completely still. A computer-generated voice. A face marked by illness, intelligence, and quiet humor. A mind traveling where the body could not: into black holes, time, gravity, singularities, and the beginning of the universe.

Stephen Hawking became one of the most recognized scientists in the world not only because of his science, but because of the contrast between his physical condition and the vastness of his thought.

His body was trapped.

His mind moved through the cosmos.

He asked questions that had once belonged mostly to religion and philosophy.

Did the universe have a beginning?

What is time?

What happens inside black holes?

Did God create the universe?

Can the laws of physics explain why there is a universe?

These questions touched the deepest human curiosity. They were not small technical questions. They were cosmic questions, the kind that make people look at the night sky and feel both wonder and fear.

Hawking did not end the debate about God.

No honest book should claim that he did.

He did not prove, in a simple final way, that no God can exist. He did not answer every question about why there are laws of physics. He did not explain absolute philosophical nothingness in a way that satisfies every philosopher.

But Hawking did something extremely important.

He made the origin of the universe a scientific question for millions of ordinary readers.

He showed that the beginning of the universe does not automatically belong to theology.

He challenged the old picture of God as the one who lights the cosmic fuse.

He argued that physical law may be enough.

That is why Hawking matters in this book.

He did not remove every mystery.

But he made God less necessary.

Why Hawking Became So Important

Stephen Hawking's fame came from several sources.

He was a brilliant theoretical physicist. He worked on black holes, singularities, general relativity, quantum theory, and cosmology. He also had a rare ability to make the deepest questions of physics feel urgent to the general public.

His book *A Brief History of Time* became a worldwide phenomenon. Many people bought it. Fewer may have fully understood every part of it. But the book changed public imagination. It brought questions about time, the Big Bang, black holes, and God into ordinary homes.

This matters.

For centuries, religion had spoken confidently about creation. The universe existed because God created it. Time began because God willed it. The laws existed because God gave them. Human beings were here because God intended them.

Hawking helped bring a different voice into public discussion.

Perhaps the universe can be understood through physics.

Perhaps time itself has a natural explanation.

Perhaps the beginning is not a supernatural moment.

Perhaps the question “What happened before the universe?” is wrongly framed.

Perhaps the universe does not need God to get started.

This was not a small cultural change.

Hawking did not merely write about science. He changed the emotional ownership of cosmic origin. He helped move creation from the pulpit to physics.

That is exactly the movement this series is tracing.

God once owned the unknown.

Science keeps taking it back.

Black Holes and the Deep Structure of Reality

Hawking is closely associated with black holes.

A black hole is a region of space-time where gravity is so strong that, beyond a boundary called the event horizon, nothing can escape, not even light.

Black holes sound like science fiction, but they are scientific objects. They arise naturally from general relativity. They are connected to massive stars, gravity, collapse, and space-time. Today, strong evidence supports their existence.

Hawking’s work on black holes was groundbreaking because he brought together ideas from general relativity, quantum theory, and thermodynamics. He showed that black holes may not be completely black. They can emit radiation, now called Hawking radiation, because of quantum effects near the event horizon.

This was startling.

Black holes had seemed like final prisons of gravity. Hawking showed that they had temperature, entropy, and quantum behavior. This opened deep questions about information, time, and the foundations of physics.

Why does this matter for God?

Because it shows again that the universe is stranger than ancient religion imagined, but its strangeness is natural.

Religion often explains mystery by turning it into divine action. Science does something more difficult and more fruitful. It studies the mystery. It builds mathematics around it. It makes predictions. It looks for evidence. It accepts correction.

A black hole is not a mouth of hell.

It is not a divine punishment.

It is not a mythic abyss.

It is a physical reality.

Hawking helped show that even the strangest parts of the universe may be approached through natural law.

Singularities Are Not Proof of God

A singularity is a place where our current physical theories break down. In general relativity, singularities appear under certain conditions, such as inside black holes and, in earlier models, at the beginning of the universe.

The word can be misleading.

It does not necessarily mean that science has found a literal point of infinite density in a final physical sense. It may mean that the theory we are using has reached its limit and can no longer describe reality properly.

This is important.

When people hear that the universe began in a singularity, they may imagine a tiny point created by God. But that is not careful science.

A singularity may signal incomplete physics.

It may show that general relativity, by itself, cannot describe the earliest universe. At very high densities and tiny scales, quantum effects likely matter. A deeper theory of quantum gravity may be needed.

This is where Hawking's work was important. He studied singularities and cosmic beginnings, but he also understood that the earliest universe required more than classical general relativity.

This should teach humility.

When science reaches a limit, religion often says, “There is God.”

But science says, “Our theory is incomplete. We need a better theory.”

That difference is essential.

A singularity is not proof of God.

It is a sign that physics has reached a frontier.

Frontiers are not altars.

They are places for further investigation.

The No-Boundary Proposal

One of Hawking’s most famous ideas, developed with James Hartle, is the no-boundary proposal.

The mathematics is complex, and this book does not need to explain it in technical detail. The general idea is that the universe may be finite but without a boundary in time in the ordinary sense.

Hawking used an analogy involving Earth.

The surface of Earth is finite. If you travel north, you can reach the North Pole. But the North Pole is not an edge where Earth’s surface suddenly begins. You cannot go north of the North Pole in the same way. The question stops making sense.

Similarly, the no-boundary idea suggests that asking what happened before the universe may be like asking what is north of the North Pole.

This does not prove that God does not exist.

It does not solve every problem.

It is a scientific proposal, and like scientific proposals, it can be debated, refined, challenged, or revised.

But its philosophical importance is great.

It challenges the simple religious picture.

Religion often imagines a clear beginning: first God, then creation. First nothing, then universe. First divine will, then space and time.

The no-boundary proposal suggests that the universe may not have the kind of sharp temporal edge that demands an external starter.

If there is no boundary in time in the ordinary sense, then the question of God lighting the cosmic fuse becomes less natural.

God as the starter of time becomes less necessary.

That is the point.

Imaginary Time and the Limits of Ordinary Language

Hawking also used the concept of imaginary time.

This phrase sounds mysterious, almost poetic. But in physics and mathematics, imaginary numbers are not fantasies. They are mathematical tools that can be useful in describing reality. Imaginary time appears in certain approaches to quantum cosmology.

For general readers, the exact mathematics is not necessary. The main point is that time near the earliest universe may not behave like ordinary clock time.

In ordinary life, time has a clear direction. Yesterday comes before today. Cause comes before effect. A beginning happens after a previous state. This is the kind of time our minds understand.

But in the deepest physics of the early universe, time may be more subtle. It may be connected to geometry, quantum states, and mathematical descriptions that do not fit ordinary intuition.

Hawking's use of imaginary time was one way to explore a universe without a sharp beginning in ordinary time.

Again, this does not give a final answer to everything.

But it weakens common-sense theology.

If time itself is not what ordinary intuition assumes, then arguments based on ordinary time become less secure. If the beginning of the universe is not like the beginning of a movie, a house, or a day, then the need for a cosmic maker is less obvious.

The universe is not required to fit the mental pictures of ancient religion.

It may be far stranger.

And far more natural.

“What Place, Then, for a Creator?”

At the end of *A Brief History of Time*, Hawking famously asked what place there would be for a creator if we found a complete theory.

This question became widely discussed because it touched both science and theology.

If the universe can be fully described by physical law, what work remains for God?

This is the same question that has followed every scientific advance.

If lightning is electricity, what work remains for the storm god?

If disease is microbial, what work remains for demons?

If species evolve by natural selection, what work remains for special creation?

If planetary motion follows gravity, what work remains for angels pushing planets?

If the universe can be described by a complete physical theory, what work remains for a creator?

The answer is uncomfortable for religion.

God becomes unnecessary.

A believer may still say that God created the laws, or that God is the reason the laws exist. But this moves God further back. God is no longer explaining the physical process. God is placed behind it.

That is a retreat, not an explanation.

Hawking's question matters because it reveals the problem of a God who only lives where science has not yet reached.

What happens if science reaches further?

History suggests the answer.

God retreats again.

The Grand Design and the Controversy Over Nothing

Later, in *The Grand Design*, Hawking and Leonard Mlodinow made a stronger public claim. They argued that because there is a law such as gravity, the universe can create itself from nothing.

This sentence became famous.

It also became controversial.

Some religious critics said that Hawking had not explained why there is gravity. Others said that his "nothing" was not absolute nothing. Others argued that laws do not create anything. Philosophers and theologians objected that physics cannot replace metaphysics.

Some of these criticisms are not entirely foolish.

We should not overstate Hawking's claim. Physics "nothing" is not always the same as philosophical absolute nothing. The existence of laws remains a deep question. The origin of the universe is not fully solved. Hawking did not end all philosophical debate.

But the larger point remains powerful.

Hawking was arguing that the universe does not need divine intervention in order to begin. He was challenging the old idea that God must start the cosmic process. He was saying that physical law may be sufficient, or at least far more explanatory than the statement "God did it."

That is the point relevant to this book.

Even if Hawking's wording is debated, his challenge to the creator hypothesis is serious.

The believer says: God must start the universe.

Hawking replies: perhaps law is enough.

Science-based atheism does not need to claim that Hawking solved every issue. It only needs to recognize that God is not required by the science.

What Hawking Did Not Prove

A fair chapter must say what Hawking did not prove.

He did not prove that no God can exist in any possible philosophical sense.

He did not prove that science has answered why there is something rather than nothing.

He did not prove that the laws of physics explain their own existence in a final way.

He did not prove that every metaphysical question is meaningless.

He did not prove that the multiverse is established fact.

He did not prove that all religious people are ignorant.

He did not prove that the universe is emotionally easy to understand without God.

These limits matter.

If we exaggerate Hawking, the argument becomes weaker. Religious critics can then attack the exaggeration instead of confronting the stronger point.

The stronger point is this:

Hawking showed that the origin of the universe can be treated as a problem in physics. He challenged the need for God as cosmic starter. He helped make clear that scientific models may go much further than religious intuition once assumed.

That is enough.

Science-based atheism should be confident, but not careless.

The honest position is stronger than the exaggerated one.

Hawking did not prove everything.

But he made God less necessary.

What Hawking Did Show

Hawking showed that the deepest cosmic questions can be approached scientifically.

He showed that black holes, time, singularities, and cosmic origins are not beyond investigation.

He helped connect general relativity, quantum theory, thermodynamics, and cosmology.

He showed that common-sense ideas of beginning may fail near the origin of the universe.

He showed that the universe may be self-contained in ways that weaken the need for an external creator.

He showed that God is not required as a working scientific hypothesis in cosmology.

This is extremely important.

For centuries, religion claimed the beginning. It claimed the origin. It claimed the cosmic first moment as divine territory. The beginning of everything seemed like the one place science could never enter.

Hawking helped science enter.

Not completely.

Not finally.

But powerfully.

He made it possible for ordinary readers to see that the question of creation is not automatically a religious question.

It is also, and perhaps primarily, a scientific question.

That is a major victory for science-based atheism.

God as the Cosmic Starter

Many religious people imagine God as the cosmic starter.

The universe is like a machine.

God turns it on.

The universe is like a fire.

God lights it.

The universe is like a story.

God writes the first sentence.

The universe is like a building.

God lays the foundation.

These images are easy to understand because they come from ordinary human experience. Machines need operators. Fires need ignition. Stories need writers. Buildings need builders.

But the universe is not a machine sitting in a workshop.

It is not a fire in a fireplace.

It is not a book on a table.

It is not a house on land.

It is the total physical reality we are trying to explain.

Hawking's work challenges these simple analogies. If space and time are part of the universe, then God does not start the universe in the way a person starts a car. If the earliest universe can be described by physical law, then no divine finger is needed to press a cosmic button.

God as cosmic starter may feel intuitive.

But intuition is not evidence.

Modern cosmology shows that the beginning question is far more subtle than religious images suggest.

Law Without Lawgiver

One common religious reply is that laws require a lawgiver.

The universe has laws.

Therefore, God must have given the laws.

This argument depends on confusion between human laws and physical laws.

Human laws are commands created by minds. Traffic laws, tax laws, criminal laws, and civil laws are written, debated, passed, enforced, and changed by human beings.

Physical laws are different.

They are not commands written on cosmic paper. They are descriptions of regularities in nature. When scientists say gravity obeys a law, they do not mean that gravity read a command and decided to comply. They mean that the behavior of matter, energy, and space-time can be described mathematically.

A physical law does not require a lawgiver in the way a traffic law does.

The word law is partly metaphorical.

This is important.

When religion says "laws need a lawgiver," it smuggles human legal language into physics. It treats the universe as if it were a kingdom and God as the king issuing decrees.

But physics does not work that way.

The laws of physics are our best descriptions of how reality behaves.

They are not evidence of a divine legislator.

Hawking's focus on law does not prove God. It weakens God by showing that natural regularity may be enough for explanation.

If There Is Law, Why God?

Suppose we accept that the universe behaves according to law.

What follows?

The believer says: God created the laws.

But again, what does this explain?

Why these laws?

Why not other laws?

How does God create laws?

Did God choose them?

Could God have chosen differently?

If God is perfect, why create laws that produce so much suffering, extinction, cosmic violence, and waste?

If laws are created by God, why do they operate with complete indifference to prayer, worship, belief, and moral worth?

Gravity pulls saint and criminal alike.

Radiation kills believer and atheist alike.

Disease strikes the devout and the skeptical.

Black holes do not care about prayer.

The laws of physics behave exactly as if they are impersonal.

This does not disprove God in a strict logical sense. A believer can always say God chooses to work through impersonal law. But that again moves God behind the evidence.

The evidence itself shows law-like regularity.

It does not show a lawgiver.

Science-based atheism accepts the laws as descriptions of reality and asks how far they can take us. Religion adds a lawgiver without evidence.

The addition is unnecessary.

The Courage of Meaning Without God

Hawking's atheistic or non-theistic statements carried emotional power because they came from a person who lived close to death for decades.

He was diagnosed with a devastating motor neuron disease as a young man. Many expected him to live only a short time. Instead, he lived for decades, worked, wrote, traveled, spoke, and became one of the most famous scientists in the world.

He did not need religion to give his life meaning.

He did not need God to think deeply.

He did not need heaven to make his time valuable.

His life became an example of courage without supernatural dependence.

This is important, though we should not use biography as proof. A person's courage does not prove atheism true. But it does show that a godless worldview need not be empty, weak, or despairing.

Hawking faced severe physical limitation and still pursued the largest questions.

He found meaning in understanding.

He found dignity in inquiry.

He found wonder in reality.

That is a powerful image for science-based atheism.

The universe does not need God to be awe-inspiring.

Human life does not need God to be brave.

The Religious Reply

A thoughtful believer may say:

"Hawking was brilliant, but he did not disprove God. His claim that the universe can create itself because of gravity still leaves the question of where gravity came from. Physical laws cannot explain their own existence. Science may describe the universe, but God explains why there is a universe at all."

This is a serious reply.

It is true that Hawking did not solve every ultimate question. It is fair to ask why there are laws, why there is reality, and why anything exists. These questions remain difficult.

But the reply still does not prove God.

Saying "God explains the laws" gives no mechanism. It gives no evidence. It gives no testable model. It does not explain how a timeless, spaceless, immaterial mind creates gravity or quantum law. It does not explain why God exists. It does not identify the God of any religion.

The question "Why is there law?" is difficult.

But “God made the laws” is not a scientific answer.

It is an assertion placed beyond testing.

Hawking may not have answered everything.

But theology has not answered everything either.

The difference is that science continues to investigate. Theology often declares victory at the boundary of knowledge.

That is not enough.

Evidence-Based Response

The evidence-based response is careful.

Stephen Hawking did not prove that no God can exist. He did not solve every problem about the origin of existence. He did not remove all mystery from cosmology.

But he showed that the universe can be approached through physical law even at its deepest boundaries. His work on black holes, singularities, quantum cosmology, and the no-boundary proposal challenged the need for God as a cosmic starter.

He helped show that the beginning of the universe is not automatically a theological event.

It may be a physical question.

That is the key point.

God is not required by Hawking's cosmology.

God may be added by belief, but God adds no scientific explanation.

Science-based atheism can accept that many questions remain while still rejecting God as unnecessary.

Hawking's contribution was not that he ended mystery.

It was that he showed mystery does not belong to God.

Chapter Summary

Stephen Hawking became one of the most famous scientists in the world because he brought deep cosmic questions to the public: black holes, time, singularities, the origin of the universe, and the role of God.

His work did not disprove God in every possible philosophical sense. He did not explain every ultimate mystery. But he powerfully challenged the idea that God is needed to start the universe.

Hawking's work on black holes, singularities, quantum cosmology, and the no-boundary proposal showed that the deepest questions of cosmic origin can be approached scientifically. His later argument that physical law may be sufficient to explain the universe weakened the old image of God as the cosmic fuse-lighter.

Hawking did not remove every mystery.

But he made God less necessary.

Key Points

Stephen Hawking brought cosmic origin questions into public discussion.

His work on black holes connected gravity, quantum theory, and thermodynamics.

Singularities may show limits of current theory, not proof of God.

The no-boundary proposal challenges simple ideas of a first moment.

Imaginary time suggests that time near the earliest universe may not behave like ordinary time.

Hawking did not prove that no God can exist.

He did challenge the need for God as cosmic starter.

Physical laws are descriptions of nature, not human laws requiring a divine lawgiver.

God explains no scientific details of cosmology.

Mystery does not belong to religion.

Transition to the Next Chapter

Hawking showed that the universe may be approached through physical law without invoking God as creator.

But a deeper issue remains.

When people say "God created the universe," what does that actually explain? Does it give a mechanism? Does it make predictions? Does it clarify why this universe exists rather than another? Does it help science understand anything better?

Or is "God did it" simply a name placed on mystery?

In the next chapter, we will examine why "God did it" is not a real explanation.

Sources

Stephen Hawking, *A Brief History of Time*

Stephen Hawking and Leonard Mlodinow, *The Grand Design*

James Hartle and Stephen Hawking, writings on the no-boundary proposal

Kip Thorne, *Black Holes and Time Warps*

Sean Carroll, *The Big Picture*

Brian Greene, *The Fabric of the Cosmos*

Victor Stenger, *God: The Failed Hypothesis*

Stanford Encyclopedia of Philosophy, "Cosmology and Theology"

Chapter 14: Why "God Did It" Explains Nothing

The Answer That Stops the Question

The most common answer in human history may also be the least explanatory:

God did it.

A child asks, "Where did the universe come from?"

The answer comes quickly.

"God made it."

The child asks, "How?"

"By His power."

The child asks, "Why?"

"Because He wanted to."

The child asks, "Where did God come from?"

"God always existed."

The child asks, "How do we know?"

"Have faith."

This may sound like an explanation, but it is not. It is a stopping point. It tells the question to stop moving. It takes the unknown, places a sacred name over it, and asks the mind to bow.

For thousands of years, this worked. Not because it explained deeply, but because it satisfied emotionally. It gave people an answer when they had no better answer. It gave authority to mystery. It gave comfort in the face of fear. It made the universe feel intentional, watched, and morally arranged.

But when we examine the sentence carefully, it becomes weak.

"God did it."

What does that actually explain?

Does it tell us how the universe began?

Does it tell us how time arose?

Does it tell us how a timeless being acts?

Does it tell us how an immaterial mind creates matter?

Does it tell us why the laws of physics have the form they do?

Does it tell us why this universe exists rather than another?

Does it make a prediction?

Does it give a mechanism?

Does it allow testing?

Does it help cosmology advance?

No.

The sentence sounds complete because it has a subject, verb, and object.

But grammar is not explanation.

"God created the universe" may be a religious belief. It may be emotionally powerful. It may feel satisfying to those already committed to faith.

But scientifically, it leaves almost everything unexplained.

Naming Mystery Is Not Explaining Mystery

Human beings often mistake naming for explaining.

In the past, a person might see lightning and say, "God is angry." That gave the event a name and a moral meaning. But it did not explain electricity, charge separation, clouds, atmospheric conditions, or discharge.

A person might see disease and say, "A demon entered the body." That gave suffering a supernatural label. But it did not explain bacteria, viruses, immunity, transmission, infection, or treatment.

A person might see an eclipse and say, "Heaven is warning us." That gave the event emotional meaning. But it did not explain the positions of the Sun, Moon, and Earth.

The same mistake happens with the universe.

When someone says, "God made the universe," the unknown receives a name. But the name does not explain the process.

It does not tell us what happened.

It does not tell us how it happened.

It does not tell us why the evidence looks the way it does.

It does not tell us what we should observe.

It does not tell us what would count against the claim.

A name can be useful if it points toward investigation. Bacteria was once only a name, but it became useful because bacteria could be observed, cultured, tested, stained, treated, and studied.

God is different.

God is usually placed beyond observation, beyond testing, beyond measurement, beyond correction, and beyond ordinary standards of evidence.

That makes God safe from disproof in the minds of believers.

But it also makes God weak as an explanation.

An explanation that cannot be examined cannot carry much weight.

What a Real Explanation Does

A real explanation does work.

It connects facts.

It clarifies causes.

It makes sense of evidence.

It allows prediction.

It can be tested.

It can be corrected.

It helps us understand why this happened rather than something else.

Consider germ theory.

It did not merely say, "Germs cause disease." It led to handwashing, sterilization, vaccination, antibiotics, infection control, water sanitation, and public health. It explained why certain diseases spread. It explained why contaminated water was dangerous. It explained why clean surgical technique saved lives.

Consider evolution.

It did not merely say, “Life changed.” It explained adaptation, common ancestry, fossil patterns, genetic relationships, antibiotic resistance, and the branching history of life.

Consider Big Bang cosmology.

It does not merely say, “The universe was once different.” It explains cosmic expansion, redshift, background radiation, light element abundance, and large-scale structure.

These explanations do work.

Now compare “God did it.”

What does it predict?

What does it measure?

What does it correct?

What new research does it open?

What specific observation would confirm it?

What specific observation would disprove it?

If the answer is “nothing,” then it is not functioning as a scientific explanation.

It may be a belief.

It may be a comfort.

It may be a theological interpretation.

But it is not an explanation in the strong scientific sense.

Mechanism Matters

A mechanism tells us how something happens.

In medicine, saying "the patient has fever" is not enough. A physician asks why. Is it pneumonia? Urinary infection? Viral illness? Sepsis? Autoimmune disease? Drug reaction? Cancer? Heat stroke?

The cause matters because treatment depends on mechanism.

In science, the same principle applies.

If the universe exists, and someone says God created it, we must ask:

How?

How does a timeless being produce time?

How does a spaceless being produce space?

How does an immaterial being produce matter?

How does a mind without a brain produce intention?

How does divine will become physical law?

How does supernatural action connect to measurable reality?

If these questions cannot be answered, then "God created" is not an explanation of mechanism. It is only an assertion of agency.

A believer may say, "God creates by divine power."

But that does not explain anything further.

What is divine power?

How does it operate?

How can we distinguish divine power from no divine power?

How does it produce specific physical outcomes?

If every question is answered by another unexplained phrase, we have not advanced. We have only changed the words.

Science does not always know the full mechanism immediately. But it searches for it. It refines it. It tests it.

Theology often protects the absence of mechanism by calling it mystery.

That is not enough.

Prediction Matters

Prediction is one of the great strengths of science.

A scientific model does not only explain what is already known. It often tells us what else we should find.

Relativity predicted the bending of light.

Big Bang cosmology predicted a background radiation from the early universe.

Evolution predicted patterns of shared ancestry.

Germ theory predicted that controlling contamination would reduce infection.

These predictions matter because they put ideas at risk. A theory that can fail is more serious than a claim that cannot fail.

What does “God did it” predict?

If the universe were small, a believer could say God made a small universe.

If the universe is vast, a believer says God made a vast universe.

If life is common, God wanted life everywhere.

If life is rare, God made Earth special.

If the universe has a beginning, God caused it.

If the universe is eternal, God sustains it eternally.

If physical laws are simple, God loves elegance.

If physical laws are complex, God’s wisdom is beyond us.

If prayers seem answered, God intervened.

If prayers fail, God had a different plan.

This is the problem.

The God explanation can be adjusted to fit any outcome.

That makes it emotionally flexible, but scientifically empty.

An explanation that fits every possible universe explains no specific universe.

Science gains strength by risk.

Religion often survives by avoiding risk.

Evidence Matters

A claim about reality needs evidence.

If someone says there is a planet orbiting a distant star, we ask for evidence.

If someone says a new drug works, we ask for evidence.

If someone says a cancer has spread, we ask for evidence.

If someone says the universe was created by an invisible, timeless, spaceless, all-powerful mind, we should also ask for evidence.

This is not disrespect.

It is intellectual responsibility.

The bigger the claim, the stronger the evidence should be.

God is the biggest claim imaginable. God is not a small adjustment to physics. God is not a minor unknown particle. God is not a hidden planet. God is claimed to be the creator of all reality, the source of the laws, the ground of existence, the moral ruler, and often the judge of human beings.

Such a claim should require extraordinary evidence.

Instead, what is often offered?

Ancient scriptures.

Personal feelings.

Religious tradition.

Mystery.

Fine-tuning.

First cause.

Awe.

Fear of death.

Moral longing.

None of these proves God.

They may explain why people believe. They do not establish the truth of the belief.

Science-based atheism insists on the same standard everywhere:

Evidence first.

Belief later.

God Explains Everything and Therefore Nothing Clearly

A good explanation must exclude some possibilities.

If an explanation is compatible with everything, it explains very little.

Suppose a doctor says, “Your symptoms are caused by invisible energy.”

The patient asks, “Why do I have fever?”

“Invisible energy.”

“Why do I have pain?”

“Invisible energy.”

“Why did the blood test change?”

“Invisible energy.”

“Why did the medicine help?”

“Invisible energy.”

“Why did the medicine fail?”

“Invisible energy.”

This would not be good medicine.

A real diagnosis must be more specific.

The same applies to God.

If God explains a universe with life and a universe without life, a universe with suffering and a universe without suffering, a universe with laws and a universe with chaos, a universe with miracles and a universe without miracles, then God explains nothing specifically.

The claim becomes immune to evidence.

Whatever happens, God is still claimed.

That is not strength.

That is vagueness.

A useful explanation tells us why one outcome happens instead of another. God is often used to explain all outcomes equally. That is why God becomes less an explanation and more a verbal blanket placed over reality.

Science removes the blanket.

It asks what is actually happening.

Divine Intention Is Not a Model

Religious explanations often rely on intention.

God wanted the universe.

God wanted life.

God wanted human beings.

God wanted moral creatures.

God wanted souls.

God wanted relationship.

God wanted worship.

But intention alone is not a scientific model.

If a person says, "The bridge exists because the engineer wanted it," that is only a small part of the explanation. We still need design, materials, calculations, construction, support, testing, and physical principles.

Wanting is not enough.

If God intended the universe, how did that intention become reality?

What connects divine will to cosmic expansion?

What connects divine choice to quantum fields?

What connects God's purpose to gravity, radiation, matter, dark energy, and galaxy formation?

Without mechanism, divine intention remains a story, not an explanation.

There is another problem.

How do we know God's intention?

Religions disagree. One says God created for glory. Another says God created for love. Another says the world is divine play. Another says creation is a test. Another says the universe is illusion. Another says there is no creator at all.

Which divine intention is correct?

Without evidence, religious intention becomes projection.

Human beings imagine purposes and place them into the cosmos.

Science-based atheism refuses to confuse human projection with cosmic truth.

The Final God of the Gaps

The God of the gaps is one of religion's oldest habits.

Where knowledge is missing, God is inserted.

When thunder was mysterious, God was in thunder.

When disease was mysterious, God was in disease.

When species were mysterious, God was in special creation.

When consciousness was mysterious, God was in the soul.

When the universe's origin is mysterious, God is placed at the beginning.

This is the final large gap.

God has moved from the visible world to the edge of cosmology.

The problem is that this method has failed again and again. The gaps that once seemed safe became scientific fields. Weather, medicine, astronomy, geology, biology, and neuroscience all advanced. God retreated.

Now God is placed in cosmic origin, fine-tuning, and the deepest laws.

But the same warning applies.

The unknown is not owned by religion.

A gap in physics is not evidence for God.

It is a problem for physics.

History strongly suggests that mysteries should be investigated, not worshiped.

Science-based atheism is built on that lesson.

Why "God Did It" Can Stop Inquiry

One danger of "God did it" is that it can stop curiosity.

If disease is punishment, why investigate microbes?

If lightning is divine anger, why study electricity?

If species were specially created, why study evolution?

If the universe was created by God, why push deeper into cosmology?

Of course, many religious scientists have investigated nature deeply. History is complex. Some believers have argued that studying nature reveals God's creation. That should be acknowledged.

But as an explanation, "God did it" has a stopping tendency.

It gives finality before understanding.

It makes ignorance feel complete.

It turns questions into obedience.

Science does the opposite.

Science says:

Do not stop at mystery.

Do not stop at authority.

Do not stop at comfort.

Do not stop at tradition.

Do not stop at "God did it."

Keep asking.

This attitude has transformed human life. It gave us medicine, technology, astronomy, chemistry, physics, biology, and modern cosmology. It reduced suffering and expanded knowledge.

"God did it" did not do that.

Curiosity did.

Evidence did.

Science did.

God as Personal Explanation

Some philosophers argue that God is a personal explanation, not a mechanical explanation.

They say we explain human actions by referring to persons and intentions. If a window is broken, and we learn that a child threw a ball, we have a personal explanation. Similarly, perhaps the universe exists because a personal God intended it.

This is an interesting reply.

But it has a weakness.

Personal explanations work when we have evidence for the person.

If a child broke the window, we may have witnesses, fingerprints, a ball, a confession, prior behavior, or video evidence. We know children exist. We know they can throw balls. We know how physical action breaks glass.

With God, the situation is different.

We do not have independent evidence of a timeless, spaceless creator. We do not know that such a person exists. We do not know how such a person acts. We do not know how divine intention produces physical reality.

A personal explanation without evidence for the person is not strong.

It is imagination.

Also, personal explanations do not eliminate mechanism. If the child broke the window, the ball still had to strike the glass with enough force. Intention does not replace physics.

If God created the universe, there still must be some account of how divine intention connects to physical reality.

Religion rarely provides that.

The Hidden Cost of the God Explanation

There is another cost to saying “God did it.”

It often makes reality less understandable.

When science explains a phenomenon, we usually understand more. When we learn that lightning is electrical discharge, the storm becomes clearer. When we learn that infection is microbial, disease becomes clearer. When we learn that life evolves, biology becomes clearer.

But when we say God did it, what becomes clearer?

Often, nothing.

Instead, we now have to explain God.

Why does God exist?

Why does God create?

Why does God choose this universe?

Why does God hide?

Why does God allow suffering?

Why does God create through long natural processes?

Why does God create laws that operate without visible divine involvement?

Why does God create a universe that looks exactly as if no God is needed?

The God explanation is not free.

It brings many new questions.

Science-based atheism avoids this unnecessary complication. It does not add a supernatural mind when natural investigation is enough. It does not place invisible intention where evidence shows physical process.

It does not multiply mysteries.

It studies reality as reality.

That is a cleaner, stronger, and more honest approach.

The Universe Does Not Need a Verb Like "Made"

Part of the problem comes from language.

We say things are made.

A chair is made.

A house is made.

A painting is made.

A machine is made.

So we ask, "Who made the universe?"

But perhaps the question is misleading.

The universe may not be the kind of thing that is made in the way chairs, houses, paintings, and machines are made. Those things are objects inside the universe, shaped from preexisting materials by beings who already exist in space and time.

The universe is not like that.

It is not an object sitting inside a larger space, waiting for a craftsman.

It includes space and time as we understand them.

So the verb made may smuggle in the wrong picture.

It imagines a maker, materials, time, action, and result. That is how human making works. But cosmic origin may not be human making.

Science is careful because language can mislead.

Religion often takes human language too literally and projects it onto reality.

God the maker.

God the king.

God the father.

God the judge.

God the lawgiver.

God the architect.

These are human images.

They may comfort the mind.

They do not prove the universe was made by a divine person.

Science as Disciplined Patience

Science requires patience.

It does not answer everything immediately.

It does not turn every mystery into certainty.

It does not pretend that hard questions are easy.

It waits for evidence.

It builds instruments.

It refines mathematics.

It tests models.

It changes when needed.

This patience is one of its highest virtues.

Religion often offers immediate answers. Who made the universe? God. Why are we here? God's plan. What happens after death? God decides. Why is there suffering? God's mysterious purpose.

These answers may comfort, but comfort is not the same as truth.

Science-based atheism asks us to tolerate uncertainty rather than fill it with fantasy.

That is not easy.

Human beings dislike uncertainty. We want closure. We want the universe to feel personal. We want to believe someone knows the answer, someone is in control, someone has a plan.

But intellectual maturity means accepting that some questions remain open.

The honest mind says:

I do not know yet.

The religious mind too often says:

I do not know, therefore God.

This book stands with the first sentence.

The Strongest Religious Reply

A thoughtful believer may say:

"God is not meant to be a scientific mechanism. God is the ultimate personal explanation for why anything exists at all. Science explains how the universe works, but God explains why there is a universe, why there are laws, and why reality is intelligible. Asking for a mechanism misunderstands God."

This is probably the strongest reply to this chapter.

It is true that believers often do not present God as a scientific mechanism. It is also true that science and philosophy ask different kinds of questions. The question of why anything exists is deeper than many ordinary scientific questions.

But the reply still has problems.

If God is offered as an explanation, we are allowed to ask what explanatory work God does.

If God gives no mechanism, no evidence, no prediction, no testable consequence, and no way to distinguish true God claims from false God claims, then God does not explain much.

Saying God is the ultimate explanation does not make it so.

God may be defined as the answer, but definition is not demonstration.

Also, if God is beyond mechanism and beyond testability, then God cannot be used as a strong reason to reject science-based atheism. God becomes a belief placed beyond evidence.

That may be faith.

It is not knowledge.

Evidence-Based Response

The evidence-based response is clear.

"God did it" does not explain the universe in any meaningful scientific sense.

It gives no mechanism.

It gives no testable model.

It makes no clear prediction.

It does not explain why this universe exists rather than another.

It does not identify which God exists.

It does not explain how divine intention becomes physical law.

It does not answer who created God.

It does not improve cosmology.

It names mystery, but does not explain mystery.

Science-based atheism does not claim that every mystery has been solved. It claims that unknowns should remain open to investigation, not closed by theology.

That is the stronger position.

God is not required as an explanation.

And when an explanation is not required, we should not add it.

Chapter Summary

"God did it" sounds like an answer, but it often functions as a stopping point rather than a real explanation.

A real explanation should clarify mechanism, connect evidence, make predictions, allow testing, and improve understanding. Scientific explanations do this. Germ theory, evolution, relativity, and Big Bang cosmology all explain specific features of reality and open paths for further knowledge.

The God explanation does not do this. It gives no mechanism, no measurement, no testable prediction, and no clear way to distinguish one religious claim from another. It can fit any possible outcome, which means it explains no specific outcome well.

God may remain a belief for some people.

But God is not a scientific explanation of the universe.

Science-based atheism is stronger because it accepts mystery without pretending that naming the mystery is the same as understanding it.

Key Points

Naming a mystery is not the same as explaining it.

A real explanation does work.

Mechanism matters.

Prediction matters.

Evidence matters.

God can be adjusted to fit any outcome, which weakens its explanatory power.

Divine intention is not a scientific model.

“God did it” often functions as a God-of-the-gaps argument.

Personal explanations require evidence for the person.

Science requires disciplined patience.

Science-based atheism accepts mystery without turning it into God.

Transition to the Next Chapter

We have now examined the main creator arguments.

The universe is vast, but vastness does not prove God.

The Big Bang is powerful science, but not creation theology.

The first-cause argument does not reach the personal God of religion.

The question “Who created God?” exposes the weakness of special pleading.

Fine-tuning is serious, but not proof.

The anthropic principle corrects our human-centered thinking.

The multiverse is uncertain, but it shows that design is not the only possible answer.

Quantum cosmology warns us that common sense may fail at the edge of reality.

Hawking showed that the universe may be approached through law without divine ignition.

Now we can bring the argument together.

In the next chapter, we will examine the cosmos without divine management.

Sources

David Hume, *Dialogues Concerning Natural Religion*

Bertrand Russell, *Why I Am Not a Christian*

J. L. Mackie, *The Miracle of Theism*

Graham Oppy, *Arguing About Gods*

Victor Stenger, *God: The Failed Hypothesis*

Sean Carroll, *The Big Picture*

Stephen Hawking and Leonard Mlodinow, *The Grand Design*

Stanford Encyclopedia of Philosophy, "Cosmology and Theology"

Chapter 15: The Cosmos Without Divine Management

The Universe Does Not Ask Permission

The universe does not pause before it acts.

Stars are born without prayer.

Galaxies collide without warning.

Black holes form without moral purpose.

Radiation travels across space without concern.

Gravity pulls without compassion.

Atoms combine without intention.

Planets form from dust, gas, heat, impact, and time.

The universe does not ask permission from gods, priests, prophets, scriptures, temples, or human wishes. It behaves according to physical reality. It expands, cools, burns, collapses, forms, destroys, and transforms.

This is not a cold statement.

It is an honest one.

For thousands of years, human beings imagined a universe under divine management. God created. God sustained. God watched. God punished. God rewarded. God arranged events. God guided history. God placed signs in the sky. God answered prayer. God held reality together by will.

Modern cosmology gives us a different picture.

The universe is vast, ancient, dynamic, lawful, and mostly indifferent to human life. It does not look like a kingdom ruled by a visible king. It does not look like a theater built for human salvation. It does not look like a house carefully designed around us.

It looks like a natural reality.

A reality of space-time, gravity, matter, radiation, quantum fields, stars, galaxies, dark matter, dark energy, expansion, structure, and time.

This does not remove wonder.

It removes divine management.

The cosmos does not need God to move.

It does not need God to expand.

It does not need God to form stars.

It does not need God to build galaxies.

It does not need God to create heavy elements.

It does not need God to produce planets.

It does not need God to be real.

That is the central conclusion toward which this book has been moving.

The universe may remain mysterious.

But God is not required.

What This Book Has Shown

This book began with the night sky.

It asked why the universe feels like one of the strongest remaining arguments for God. Thunder no longer needs a god. Disease no longer needs demons. Planetary motion no longer needs angelic guidance. Life will be examined more fully in the next book, but even there science has already transformed the question. Yet the universe itself still feels like sacred territory.

Who made all this?

Why does anything exist?

Why are there laws?

Why is the universe life-permitting?

Did the Big Bang need a creator?

These questions are serious.

But serious questions do not automatically produce religious answers.

We have seen that the universe is not a human-centered stage. Earth is one small planet. The Sun is one ordinary star. The Milky Way is one galaxy among billions. Human beings appeared very late in cosmic history. Most of the universe is not friendly to life.

We have seen that the sacred sky became the physical cosmos. Heaven was replaced by space-time, stars, galaxies, radiation, gravity, and cosmic history.

We have seen that Einstein changed the meaning of space, time, gravity, and motion. The stage itself became natural.

We have seen that Hubble revealed an expanding universe. The cosmos became dynamic, not static.

We have seen that the Big Bang is a scientific model, not a religious creation story.

We have seen that evidence-based cosmology rests on expansion, cosmic microwave background radiation, light element abundance, and large-scale structure.

We have seen that first-cause arguments do not prove a personal God.

We have seen that "Who created God?" exposes special pleading.

We have seen that fine-tuning is a real question, but not proof of design.

We have seen that the anthropic principle corrects human-centered thinking.

We have seen that multiverse ideas, though unproven, show that design is not the only possible explanation.

We have seen that quantum cosmology warns us not to force ordinary ideas of cause, time, and nothing onto the earliest universe.

We have seen that Stephen Hawking challenged the need for God as cosmic starter.

And we have seen that "God did it" is not a real scientific explanation.

Now these threads come together.

The cosmos does not need divine management.

The Big Bang Is Not God

The Big Bang is one of the most misunderstood ideas in modern culture.

To many believers, it sounds like creation. A beginning. A first moment. A burst of existence. A place where God can enter.

But Big Bang cosmology is not theology.

It describes the observable universe expanding and cooling from an earlier hot, dense state. It is supported by evidence. It connects redshift, cosmic background radiation, light element abundance, and large-scale structure.

It does not say God created the universe.

It does not say nothing became something by divine command.

It does not prove Genesis.

It does not prove the Quran.

It does not prove any scripture.

It does not prove a personal creator.

The Big Bang gives us a physical history of the observable universe. That history is magnificent. But it is not a religious event.

Religious arguments often take the edge of scientific knowledge and place God there. They say that because the Big Bang points toward an early state, God must have caused it.

But this is a leap, not a conclusion.

A scientific boundary is not a temple.

A difficult question is not evidence for God.

The Big Bang does not prove that God exists. It proves that the universe has a natural history that can be studied through science.

That is already extraordinary.

And it is enough.

First Cause Is Not the God of Religion

The first-cause argument remains attractive because it gives the mind a stopping point.

Everything must have a cause.

The universe began.

Therefore, the universe must have a cause.

That cause is God.

But this simple chain hides serious problems.

Cause is a concept we know from events inside the universe. The universe as a whole is not just another object inside a larger known space. If time itself is part of the universe, then the idea of a cause before the universe becomes difficult.

Even if some ultimate cause or foundation exists, it does not automatically become the God of religion.

It does not become a loving God.

It does not become a moral God.

It does not become a prayer-answering God.

It does not become a God who writes scriptures.

It does not become a God who judges sin.

It does not become a God who promises heaven or threatens hell.

It does not become the God of any particular religion.

At most, the first-cause argument points to a mystery about ultimate reality. Religion then adds personality, intention, morality, worship, revelation, and authority. But those additions are not established by the argument.

This is one of the most important points in the whole book.

A first cause, even if granted, is not enough for religion.

The road from "the universe has an explanation" to "my God is real" is far longer and weaker than believers often admit.

Science-based atheism does not need to answer every ultimate question to reject that leap.

It only needs to show that the leap is not justified.

Fine-Tuning Is Not Proof

Fine-tuning may be the strongest modern argument for God.

It deserves respect because it takes science seriously. It asks why the universe has features that allow stars, atoms, chemistry, planets, and life. It recognizes that certain physical values appear important for complexity.

But fine-tuning is not proof of God.

We do not know whether the constants could vary freely.

We do not know the full range of possible universes.

We do not know the probability distribution of possible laws and constants.

We do not know whether deeper physics will explain what now looks finely tuned.

We do not know whether many cosmic domains exist.

We do not know whether other forms of life could exist under different conditions.

And we do know that the universe is not broadly friendly to life.

Most of it is empty, dark, freezing, burning, irradiated, airless, or lethal. Even Earth is full of danger, disease, extinction, and suffering. A universe designed lovingly for human life would be expected to look very different from the one we observe.

Fine-tuning raises questions.

It does not answer them with God.

A mystery in physics does not become evidence for theology. The design argument may feel emotionally powerful because human beings naturally infer mind from order. But the universe is not a watch, a house, or a machine made inside a workshop. It is the whole physical reality we are trying to understand.

Science-based atheism can accept fine-tuning as a serious issue without surrendering to design.

A serious mystery is still not a divine signature.

The Anthropic Correction

The anthropic principle helps correct our perspective.

We observe a universe compatible with observers because only such a universe could contain observers.

This sounds simple, but it matters deeply.

We should not be shocked to find ourselves in conditions where beings like us can exist. If those conditions did not exist, we would not be here asking questions.

This does not solve every mystery.

It does not explain why the laws exist.

It does not explain why there is something rather than nothing.

It does not prove a multiverse.

It does not answer all fine-tuning questions.

But it weakens a common religious mistake.

Religion often says: the universe fits us, therefore it was made for us.

Science replies: we fit the universe because we arose within it.

That is a different way of seeing reality.

We are not outside the universe, inspecting it as a finished product. We are part of the universe. We are made from its matter, shaped by its laws, and limited by its conditions. We can only observe the kind of reality that permits observers.

This does not make life meaningless.

It makes life natural.

We are not the purpose of the cosmos.

We are one of its outcomes.

That is not an insult. It is a truth large enough to hold wonder.

Multiverse and Humility

Multiverse ideas are uncertain.

They should not be exaggerated. They should not be treated as established fact. They should not become atheist dogma.

But they matter.

They show that divine design is not the only possible response to fine-tuning. If many cosmic regions exist, perhaps with different conditions, then observers would naturally arise only in those regions where observers can arise. We would not find ourselves in sterile universes because no one would be there to observe them.

This possibility changes the design argument.

The believer says: one life-permitting universe is too unlikely, therefore God.

Science can respond: perhaps there are many cosmic domains, deeper laws, selection effects, or natural explanations we do not yet understand.

The point is not that the multiverse is proven.

The point is that God is not forced.

This is how science-based atheism should argue: not by pretending to know more than it knows, but by refusing unsupported religious certainty.

Scientific humility is not weakness.

It is strength.

Religion often claims final answers before evidence is available. Science says, “We do not yet know.” That sentence is not failure. It is honesty.

The unknown belongs to inquiry, not to God.

Quantum Cosmology and the Limits of Common Sense

Quantum cosmology teaches caution.

At the deepest levels, ordinary words may fail.

Before.

After.

Cause.

Beginning.

Nothing.

These words work well in daily life, but the earliest universe is not daily life. Time may not behave as ordinary intuition suggests. Cause may not work in a simple before-and-after way. Physics "nothing" is not the same as absolute philosophical nothingness. A boundary in our present knowledge is not proof of divine creation.

This matters because many religious arguments depend on common sense.

Everything that begins must have a cause.

Nothing can come from nothing.

A law must have a lawgiver.

A universe must have a maker.

These claims feel obvious because they fit ordinary life. But modern physics has repeatedly shown that ordinary intuition often fails at cosmic and quantum scales.

Earth feels still, but it moves.

Time feels absolute, but it is not.

Matter feels solid, but its deeper structure is strange.

Light behaves in ways ordinary sense did not predict.

Gravity is not simply a pull through empty space, but is related to space-time curvature.

The universe does not owe us intuitive simplicity.

When common sense fails, God should not be inserted into the confusion.

The honest response is further investigation.

Hawking and the End of Cosmic Ignition

Stephen Hawking helped millions see that cosmic origin can be approached scientifically.

He did not prove that no God can exist.

He did not solve every philosophical question.

He did not fully explain why laws exist.

He did not make all mysteries disappear.

But he powerfully challenged the idea of God as the cosmic starter.

The old religious picture is simple: first God, then universe. God lights the fuse. God starts the clock. God speaks, and reality appears.

Hawking's work made this picture less necessary.

Black holes, singularities, quantum cosmology, imaginary time, and the no-boundary proposal all suggest that the deepest structure of reality may not fit the human image of a maker beginning a machine.

The universe may be self-contained in ways we do not yet fully understand.

The beginning may not be a sharp edge requiring divine ignition.

Physical law may go further than theology expected.

Again, caution is necessary. Hawking should not be exaggerated. But neither should his importance be minimized. He helped move cosmic origin away from religious imagination and into scientific thought.

That is a major step in the retreat of God.

Why Divine Management Is Unnecessary

A universe under divine management should show signs of divine management.

But what do we observe?

We observe physical regularity.

We observe gravity acting without moral preference.

We observe radiation traveling without spiritual intention.

We observe galaxies forming through natural processes.

We observe stars living and dying through physics.

We observe planets forming from dust and gas.

We observe cosmic expansion.

We observe laws that operate whether or not anyone prays.

We observe a universe indifferent to human belief.

The laws of nature do not change for the devout.

Gravity does not weaken for the righteous.

Cancer does not avoid the prayerful.

Earthquakes do not spare temples.

Stars do not burn according to scripture.

Black holes do not obey theology.

The universe behaves as if it is natural.

Religion may say God manages everything invisibly through natural law. But if divine management is indistinguishable from no divine management, then it has no explanatory value.

Science explains what we observe through natural processes.

God adds nothing necessary.

That is the point.

A God who is never needed to explain the evidence becomes unnecessary as an explanation.

The Universe Is Lawful, Not Moral

One of the most important distinctions is this:

The universe is lawful, but not moral.

Gravity behaves regularly, but it does not care.

Fire burns the innocent and guilty alike.

Radiation damages tissue without asking about virtue.

A planet can support life or destroy it without moral intention.

A supernova can create elements and destroy worlds.

The universe has structure, but not justice.

This is very different from the religious picture.

Religion often imagines the universe as morally governed. God rewards, punishes, guides, teaches, tests, and judges. Cosmic events are tied to moral meaning. Disaster becomes warning. Disease becomes punishment or trial. Good fortune becomes blessing. Survival becomes grace.

Science removed this moral reading from nature.

The cosmos is not a courtroom.

It is not a church.

It is not a moral machine.

It is a physical reality.

This does not mean morality is unreal. It means morality arises within conscious beings, not from galaxies, gravity, or cosmic expansion. Moral responsibility belongs to us. We cannot outsource it to the universe.

Science-based atheism does not deny morality.

It removes false cosmic moralism.

That is progress.

A Universe Without God Is Not Empty

Many people fear that a universe without divine management is empty.

They imagine silence, loneliness, and meaninglessness.

But this fear comes from expecting the universe to behave like a parent.

The universe is not a parent.

It does not love us.

It does not hate us.

It does not guide us.

It does not punish us.

It does not promise reunion after death.

But that does not make life meaningless.

Meaning is not something that must fall from the sky. Meaning is created and experienced by conscious beings. Love matters because we love. Pain matters because we suffer. Truth matters because minds can seek it. Justice matters because lives can be harmed or protected. Beauty matters because conscious beings can perceive and cherish it.

A godless universe is not empty.

It contains stars, galaxies, planets, life, minds, love, suffering, art, courage, knowledge, and compassion.

It contains us.

That is enough to make responsibility urgent.

If no divine manager is arranging the world, then human beings must take greater responsibility. We must protect life, reduce suffering, seek truth, build justice, care for Earth, and comfort one another.

Atheism does not remove meaning.

It removes illusion.

Science-Based Atheism and Cosmic Maturity

Science-based atheism is not merely saying no to God.

It is saying yes to reality.

Yes to evidence.

Yes to reason.

Yes to humility.

Yes to uncertainty where uncertainty remains.

Yes to wonder without worship.

Yes to meaning without cosmic flattery.

Yes to morality without divine command.

Yes to truth even when truth is uncomfortable.

This is what makes it mature.

Religion often keeps humanity in a childlike relationship with the universe. A father watches. A father protects. A father punishes. A father forgives. A father has a plan. A father will make everything right.

Science-based atheism asks us to grow up.

There may be no cosmic father.

There may be no hidden plan.

There may be no final rescue from outside nature.

There may be no divine explanation behind the stars.

But there is reality.

And reality can be understood.

Not completely yet. Perhaps never completely. But enough to live more honestly than ancient religion allowed.

The universe without divine management is not a loss.

It is intellectual adulthood.

The Strongest Religious Reply

A thoughtful believer may say:

"Science explains the processes of the universe, but it does not explain why those processes exist. God may not be needed to calculate gravity or cosmic expansion, but God is needed to explain why there is a lawful universe at all. God sustains the laws, gives reality its existence, and provides ultimate meaning."

This is the strongest religious reply.

It is more subtle than saying God directly pushes planets or lights stars. It accepts science but places God behind the scientific order.

But the reply still has problems.

First, saying God sustains the laws gives no testable evidence.

Second, it does not explain how God sustains anything.

Third, it does not explain why God exists.

Fourth, it does not show that the ultimate foundation is personal.

Fifth, it does not prove any religion.

Sixth, it does not improve scientific understanding.

Seventh, it may simply rename the mystery of existence.

A believer may choose to interpret reality through God.

But the interpretation is not required.

The universe can be studied, understood, and valued without adding divine management. The laws work without visible divine enforcement. The cosmos evolves without supernatural intervention. Meaning can arise within conscious life.

God remains optional belief, not necessary explanation.

That is the central answer.

Evidence-Based Response

The evidence-based response is clear.

The cosmos does not require divine management to explain what we observe.

Modern cosmology explains the universe through expansion, radiation, gravity, space-time, matter, energy, quantum fields, stars, galaxies, dark matter, dark energy, and cosmic evolution. It does not answer every ultimate question, but it explains vastly more than religion ever did.

God adds no scientific mechanism.

God makes no testable prediction.

God does not clarify cosmic expansion.

God does not explain the cosmic microwave background.

God does not solve dark matter or dark energy.

God does not explain why the constants have their values.

God does not show why this universe exists rather than another.

God does not identify which religion is true.

God can be added as belief, but not as necessity.

Science-based atheism is the stronger position because it accepts what evidence shows and does not add what evidence does not require.

The universe remains mysterious.

But mystery is not God.

Chapter Summary

This chapter brought together the central argument of Book 3.

The universe is vast, ancient, dynamic, lawful, and deeply mysterious. But none of this proves God. The Big Bang is not a religious creation story. The first-cause argument does not prove a personal God. The question "Who created God?" exposes special pleading. Fine-tuning is serious but not proof. The anthropic principle corrects human-centered thinking. Multiverse ideas show that design is not the only possible explanation. Quantum cosmology challenges common-sense creator arguments. Hawking showed that cosmic origin can be approached through physical law.

God is not needed to manage the cosmos.

The universe behaves according to natural processes. Divine management adds no scientific explanation. God may remain a belief for some people, but God is not required by cosmology.

Science-based atheism accepts the universe as it is: magnificent, mysterious, natural, and real.

Key Points

The universe behaves according to natural processes, not visible divine management.

The Big Bang is not God.

A first cause is not automatically a personal God.

Fine-tuning is not proof of divine design.

The anthropic principle weakens human-centered interpretation.

The multiverse is not proven, but it shows design is not forced.

Quantum cosmology warns that common sense may fail near cosmic origin.

Hawking challenged God as cosmic starter.

The universe is lawful, but not moral.

A godless universe can still contain meaning, value, beauty, love, and responsibility.

Science-based atheism is intellectual maturity, not despair.

Transition to the Conclusion

We have now reached the central conclusion of this book.

The universe does not need God as creator, lawgiver, designer, first cause, or hidden manager of cosmic processes.

Still, mystery remains.

Why is there something rather than nothing? Why do the laws exist? Why does the universe have the structure it has? Why is reality intelligible at all?

These are real questions.

But the lesson of this book is clear: unanswered questions are not evidence for God.

In the conclusion, we will gather the full argument and return to the largest idea of all:

A universe without a creator is not a universe without wonder.

It is a universe large enough for truth.

Sources

Stephen Hawking, *A Brief History of Time*

Stephen Hawking and Leonard Mlodinow, *The Grand Design*

Sean Carroll, *The Big Picture*

Steven Weinberg, *The First Three Minutes*

Brian Greene, *The Fabric of the Cosmos*

Carl Sagan, *Cosmos*

Victor Stenger, *God: The Failed Hypothesis*

Stanford Encyclopedia of Philosophy, "Cosmology and Theology"

Conclusion: The Universe Without God

The Silence of the Stars

The universe does not speak in human language.

It does not announce a creator from the stars. It does not write scripture across the sky. It does not tell us that Earth is the center of reality. It does not promise that human beings are the purpose of all existence. It does not say that galaxies were made for us, that black holes carry moral meaning, or that cosmic expansion follows a divine plan.

Yet the universe is magnificent.

The night sky is still beautiful. Galaxies are still breathtaking. The cosmic microwave background remains one of the most astonishing discoveries in human history. Black holes still stretch imagination. Space-time is still strange. The Big Bang is still profound. The existence of matter, energy, stars, planets, life, and consciousness still fills the honest mind with wonder.

But wonder is not proof of God.

That is the central lesson of this book.

For a long time, human beings looked at the universe and saw divine intention. The sky looked like heaven. The stars looked like sacred lamps. Earth felt central. The universe seemed like a stage for human life, sin, prayer, judgment, and salvation.

Modern cosmology changed that.

The universe became vast. Earth became small. The Sun became one star among many. The Milky Way became one galaxy among billions. Space and time became physical. Gravity became geometry. The universe began to expand. The early universe left radiation behind. Cosmic history became measurable.

God did not become more necessary.

God became less necessary.

Not because every mystery disappeared, but because the mysteries became scientific and philosophical questions, not automatic evidence for a creator.

The universe remains deep.

But depth is not deity.

The Big Bang Is Not a Divine Signature

The Big Bang has often been misunderstood.

Many people imagine it as a scientific version of creation. They picture God standing before nothingness, lighting a cosmic fuse, and producing space, time, matter, energy, stars, planets, and life.

But that is not what Big Bang cosmology says.

The Big Bang is a scientific model of the observable universe expanding and cooling from an earlier hot, dense state. It is supported by evidence: cosmic expansion, galactic redshift, cosmic microwave background radiation, light element abundance, and the large-scale structure of the universe.

This is powerful science.

It is not theology.

The Big Bang does not prove Genesis.

It does not prove the Quran.

It does not prove any scripture.

It does not prove a personal creator.

It does not prove heaven, prayer, miracles, divine judgment, or life after death.

It tells us that the observable universe has a physical history.

That history is grand enough without turning it into religion.

A beginning, even if accepted in some sense, is not automatically a God. A beginning is a question. God is one proposed answer. Proposed answers need evidence.

In this case, the evidence supports cosmology.

It does not support divine creation.

The First Cause Does Not Reach Religion

The first-cause argument has emotional force.

It asks: if everything has a cause, what caused the universe?

But the argument becomes weaker when examined carefully.

Cause is something we understand from events inside the universe. One thing happens, then another follows. But the universe is not simply one object inside a larger known space. It includes space and time as we understand them. If time itself is part of the universe, then the phrase before the universe becomes difficult.

Even if there is some ultimate cause, foundation, or explanation of reality, that does not automatically prove the God of religion.

A first cause is not necessarily personal.

It is not necessarily loving.

It is not necessarily conscious.

It is not necessarily moral.

It does not hear prayer.

It does not write scripture.

It does not promise heaven.

It does not judge sin.

It does not prove any religion true.

This is where many religious arguments overreach. They begin with a philosophical question and end with the God they already worship. But the bridge between the two is weak.

The universe may have a deep foundation.

But calling that foundation God does not make religion true.

Fine-Tuning Is a Question, Not a Proof

Fine-tuning is one of the strongest modern arguments for God, and it should be treated seriously.

The universe appears to have features that allow stars, atoms, chemistry, planets, and life. If certain constants or conditions were very different, life as we know it might not exist.

This is interesting.

It may be profound.

But it is not proof of God.

We do not know whether the constants could have varied freely. We do not know the full range of possible universes. We do not know the probability distribution. We do not know whether deeper physical laws determine the values. We do not know whether many cosmic domains exist. We do not know whether different forms of life could exist under different conditions.

And we do know this:

The universe is not broadly friendly to life.

Most of it is empty, frozen, burning, irradiated, airless, or lethal. Even Earth is full of disease, disaster, pain, extinction, and death. A universe designed mainly for human life is not the most obvious reading of the evidence.

Fine-tuning raises a real question.

But God is not the answer forced by that question.

A mystery in physics remains a mystery in physics. It does not become a divine signature merely because we do not yet know the final explanation.

The Anthropic Lesson

The anthropic principle gives us a necessary correction.

We observe a universe compatible with observers because only such a universe could contain observers.

This does not answer everything. It does not solve all questions about fine-tuning. It does not explain why there are laws, why there is reality, or why there is something rather than nothing.

But it corrects a serious human mistake.

We should not be surprised to find ourselves in conditions where beings like us can exist. If those conditions were absent, we would not be here to ask questions.

Religion often says:

The universe fits us, therefore it was made for us.

Science replies:

We fit this universe because we arose within it.

That is a very different interpretation.

It is less flattering.

But it is more careful.

We are not outside the universe, evaluating it as a product made for our use. We are inside it. We are made of its matter. We are shaped by its laws. We are possible only because its conditions allowed beings like us to develop.

The universe may not be made for us.

We may simply be one outcome of it.

The Unknown Is Not Owned by God

This book has admitted uncertainty.

That is important.

Science has not answered every question about the universe. It has not fully explained dark matter. It has not fully explained dark energy. It has not completed quantum gravity. It has not solved every question about cosmic inflation. It has not given a final answer to why there is something rather than nothing.

But none of this proves God.

The unknown is not owned by religion.

This may be the most important intellectual discipline in the entire book.

When we do not know, the honest answer is:

We do not yet know.

The dishonest answer is:

We do not know, therefore God.

History has repeatedly punished that mistake. God was placed in thunder, disease, planetary motion, special creation, deep time, and the soul. Science entered those spaces and gave better explanations. God retreated.

Now God is placed in cosmic origin, fine-tuning, and the laws of physics.

But the lesson should be clear.

A gap in knowledge is not evidence for God.

It is an invitation to inquiry.

Hawking and the Scientific Ownership of Cosmic Origin

Stephen Hawking did not solve every mystery.

He did not disprove every possible idea of God. He did not answer every philosophical question. He did not explain absolute nothingness in a way that satisfies everyone.

But he did something historically important.

He helped show that cosmic origin can be approached scientifically.

Black holes, singularities, imaginary time, quantum cosmology, and the no-boundary proposal all point to the same broad lesson: the beginning of the universe is not automatically a religious event.

It may be a physical question.

That alone is powerful.

For centuries, religion claimed the beginning. God was the one who started time, lit the cosmic fuse, and gave reality its first push. Hawking helped millions see that this picture may be unnecessary.

Physical law may go deeper than religious imagination expected.

God may not be needed even at the edge where religion hoped to keep Him.

Hawking did not end mystery.

He helped remove God's ownership of mystery.

Why "God Did It" Is Not Enough

The phrase "God did it" sounds like an answer.

But when examined carefully, it explains very little.

It gives no mechanism.

It gives no measurement.

It gives no prediction.

It gives no testable model.

It does not explain why this universe exists rather than another.

It does not explain how a timeless being acts.

It does not explain how an immaterial mind creates matter.

It does not explain why the laws are what they are.

It does not identify which God exists.

It does not prove any scripture.

It does not improve cosmology.

It names the mystery, but naming is not explaining.

Science does something different. It asks what happened, how it happened, what evidence supports it, what predictions follow, and what would force correction.

That is why science has advanced.

Religion often gives certainty before understanding.

Science gives understanding by refusing false certainty.

The Universe Without Divine Management

The universe behaves as if it is natural.

Gravity does not ask who is righteous.

Radiation does not avoid the devout.

Galaxies do not form by prayer.

Stars do not burn according to scripture.

Black holes do not obey theology.

Cosmic expansion does not depend on worship.

The laws of nature operate with complete indifference to human belief.

This does not prove that no God can be imagined. A believer can always say God works invisibly behind natural law. But if divine management is indistinguishable from no divine management, then it has no explanatory power.

A hidden God who changes nothing in the evidence is not needed to explain the evidence.

The cosmos can be understood as a natural reality.

Not completely yet.

Perhaps never completely.

But far more honestly than ancient religion ever allowed.

The universe does not need God as creator, lawgiver, first cause, fine-tuner, cosmic starter, or hidden manager.

That is the central conclusion of this book.

A Godless Universe Is Not a Meaningless Universe

Some readers may feel the emotional weight of this conclusion.

If there is no creator, then who planned us?

If there is no divine lawgiver, then where does meaning come from?

If the universe is not made for us, then do we matter?

These are human questions. They deserve compassion.

But the answer is not despair.

A godless universe is not meaningless.

It is real.

Meaning does not need to be written into the stars before life begins. Meaning can arise within conscious beings who love, suffer, remember, create, learn, and care. A mother's love does not become false because galaxies are ancient. A child's pain does not become unimportant because Earth is small. Justice does not become meaningless because the universe is not morally governed from above.

Meaning is not less real because it is human.

In fact, meaning becomes more urgent.

If there is no divine manager, then we must care for one another.

If there is no cosmic rescue, then we must reduce suffering here.

If there is no heavenly guarantee of justice, then justice must be built by human hands.

If there is no God watching over Earth, then Earth is our responsibility.

Science-based atheism does not remove moral responsibility.

It increases it.

Wonder Without Worship

The universe does not need God in order to inspire awe.

A star is not less magnificent because it is nuclear fusion.

A galaxy is not less beautiful because gravity shapes it.

The cosmic microwave background is not less astonishing because it is radiation.

Space-time is not less profound because it is physical.

Black holes are not less terrifying because they are natural.

The Big Bang is not less mysterious because it is cosmology.

Understanding does not kill wonder.

It deepens it.

Religion often says that without God, reality becomes cold. But this is not true. Reality becomes clearer. The false decorations fall away, and the actual universe appears in its full grandeur.

A universe without God is not smaller.

It is larger than ancient religion imagined.

It is older.

Stranger.

Deeper.

More subtle.

More powerful.

More real.

Science-based atheism does not ask us to stop wondering.

It asks us to stop worshiping our own explanations.

It asks us to let reality be greater than ancient imagination.

That is not loss.

That is liberation.

From Cosmic Childhood to Intellectual Maturity

Religion often keeps humanity in a childlike relationship with the universe.

A father watches.

A father protects.

A father punishes.

A father forgives.

A father has a plan.

A father will make everything right.

This picture is emotionally powerful because it speaks to deep human needs. We want safety. We want justice. We want love. We want someone greater than us to know what we cannot know and fix what we cannot fix.

But adulthood begins when we accept reality without demanding that it behave like a parent.

Science-based atheism is not cold rebellion.

It is intellectual maturity.

It says the universe may not have a father.

It may not have a plan.

It may not have a judge.

It may not have a savior.

But it has reality.

It has evidence.

It has law.

It has beauty.

It has minds capable of understanding.

It has living beings capable of compassion.

This is enough to build a serious life.

A mature worldview does not need comforting illusion to remain meaningful.

It needs honesty.

A Universe Large Enough for Truth

The universe is large enough for truth.

It does not need to be forced into scripture.

It does not need to fit human comfort.

It does not need to make us central.

It does not need to prove our religions.

It does not need to justify our hopes.

It does not need to protect our illusions.

The mature mind can stand under the night sky and say:

I am small.

I am temporary.

I do not know everything.

The universe was not made around me.

But I can still think.

I can still love.

I can still seek truth.

I can still reduce suffering.

I can still live with dignity.

I can still feel wonder.

That is enough.

Science-based atheism is not a philosophy of emptiness. It is a philosophy of intellectual honesty. It does not pretend that the universe has a divine author when evidence does not show one. It does not call mystery God.

It does not confuse awe with proof. It does not turn ignorance into worship.

It accepts reality as reality.

And reality is enough.

Bridge to the Next Book

This book has examined the universe.

It has asked whether the cosmos needs a creator, a first cause, a lawgiver, a designer, or a hidden divine manager. The answer has been clear: modern cosmology does not require God as an explanation.

But one of the oldest and strongest religious arguments remains.

Life.

Many people can accept that astronomy explains the stars. They can accept that physics explains gravity. They may even accept that cosmology explains the early universe. But when they look at the living world, they still feel design.

The eye.

The wing.

The cell.

DNA.

The human body.

The brain.

The beauty of animals and plants.

The complexity of life seems to cry out for a designer.

But does it?

The next book in this series will examine life without a designer.

It will explore evolution, natural selection, genetics, DNA, common ancestry, imperfect design, extinction, suffering in nature, and why life does not require God.

The universe does not need a creator.

Now we must ask whether life needs a designer.

Final Thought

The universe is not evidence that God exists.

It is evidence that reality is vast, ancient, lawful, strange, and magnificent.

The Big Bang does not prove creation.

Fine-tuning does not prove design.

A first cause does not prove a personal God.

A law does not prove a lawgiver.

A mystery does not prove the divine.

The cosmos does not need God in order to be real.

And we do not need God in order to stand beneath the stars with honesty, humility, and wonder.

The universe does not need to be created for us in order to be magnificent.

It only needs to be true.

Sources

Stephen Hawking, *A Brief History of Time*

Stephen Hawking and Leonard Mlodinow, *The Grand Design*

Sean Carroll, *The Big Picture*

Steven Weinberg, *The First Three Minutes*

Brian Greene, *The Fabric of the Cosmos*

Carl Sagan, *Cosmos*

Victor Stenger, *God: The Failed Hypothesis*

Stanford Encyclopedia of Philosophy, "Cosmology and Theology"

Appendix: Main Sources and Further Reading on Cosmology, Physics, and Creator Arguments

Why This Appendix Is Included

This book has argued that the universe does not need God as creator, first cause, lawgiver, fine-tuner, or hidden divine manager. That argument does not rest on one scientist, one observation, one philosophical objection, or one clever sentence. It rests on a broad pattern of scientific discovery and careful reasoning.

Modern cosmology is not guesswork. It is built from many lines of evidence: cosmic expansion, galactic redshift, cosmic microwave background radiation, light element abundance, large-scale structure, gravity, space-time, dark matter, dark energy, and increasingly precise measurements from telescopes, satellites, and space missions.

At the same time, philosophical arguments for God deserve serious treatment. The first-cause argument, the Kalam cosmological argument, the principle of sufficient reason, necessary being, fine-tuning, and the claim that natural laws require a lawgiver have shaped religious thought for centuries. They should not be dismissed casually. They should be understood clearly, then examined honestly.

This appendix is not meant to be a complete bibliography of cosmology, physics, philosophy, atheism, or theology. It is a selected guide to important writers, books, concepts, and sources that help readers understand the central question of this volume:

Does the universe require God?

The answer developed in this book is no.

But the road to that answer should pass through evidence, reason, humility, and intellectual honesty.

Scientific and Cosmological Foundations

Albert Einstein

Key works:
Relativity: The Special and the General Theory
Scientific papers on special relativity, general relativity, and cosmology

Einstein changed the meaning of space, time, gravity, matter, and energy. Before Einstein, space often seemed like an empty container, and time seemed like a universal clock. Einstein showed that space and time are connected, that gravity is related to the curvature of space-time, and that reality is far stranger than ordinary common sense suggests.

His work matters deeply for this book because many creator arguments depend on simple ideas of time, cause, space, and beginning. Einstein showed that those ideas must be treated carefully at cosmic scale.

Einstein did not prove atheism by himself. But his science made the universe more deeply natural and less dependent on divine management.

Edwin Hubble

Key works:
Scientific papers on galaxies, Cepheid variables, redshift, and cosmic expansion

Hubble helped show that the Milky Way is not the whole universe. By studying distant galaxies, especially Andromeda, he helped reveal that the cosmos is far larger than earlier generations imagined.

He also helped establish the relationship between galaxy distance and redshift, leading to the modern understanding that the universe is expanding.

Hubble matters because cosmic expansion changed the creator debate. It gave the observable universe a physical history. But that history belongs to evidence-based cosmology, not to religious creation stories.

Georges Lemaître

Key works:
Scientific writings on the expanding universe and the primeval atom

Lemaître was both a physicist and a Catholic priest. His work on the expanding universe and the idea of a primeval atom helped prepare the way for Big Bang cosmology.

His life is important because it prevents a simplistic story in which science and religion are always crude enemies. A religious person can do excellent science. But Lemaître's scientific work did not prove God or Catholic doctrine. His cosmology was based on physics, mathematics, and observation, not scripture.

The private faith of a scientist and the explanatory power of science are not the same thing.

Alexander Friedmann

Key works:
Scientific papers on dynamic cosmological models

Friedmann showed that Einstein's equations allowed expanding and contracting universe models. This was a major step away from the assumption of a static universe.

His work helped open the door to the modern understanding of the universe as dynamic, not fixed. A dynamic universe has a physical history, and that history can be investigated by science.

Arno Penzias and Robert Wilson

Key work:
Discovery of the cosmic microwave background radiation

Penzias and Wilson detected the cosmic microwave background radiation in 1965. This faint radiation, coming from all directions, became one of the strongest pieces of evidence for Big Bang cosmology.

Their discovery matters because it showed that the early universe left a measurable physical trace. The universe did not leave behind a divine signature.

It left radiation.

Science learned to read it.

George Gamow, Ralph Alpher, and Robert Herman

Key works:
Writings on early universe physics, light element formation, and background radiation

Gamow, Alpher, and Herman contributed to early Big Bang theory, including work related to the hot early universe, the formation of light elements, and the expected relic radiation.

Their work shows that Big Bang cosmology developed through prediction, calculation, and evidence. It was not a religious story renamed as science. It was a scientific model that could be tested.

Steven Weinberg

Key work:
The First Three Minutes

Weinberg's book remains one of the classic accounts of the early universe. It explains how modern physics reconstructs the first moments of cosmic history, especially the formation of light elements and the early hot universe.

Weinberg is important because he presents cosmology as evidence-based science. His work also reflects a naturalistic view of the universe, one that does not require divine explanation.

Stephen Hawking

Key works:
A Brief History of Time
The Grand Design
Scientific writings on black holes, singularities, and cosmology

Hawking brought cosmic questions into public life. He studied black holes, singularities, time, and the origin of the universe. His work helped millions understand that the beginning of the universe may be approached through physics rather than theology.

Hawking did not prove that no God can exist in every philosophical sense. But he powerfully challenged the need for God as the cosmic starter.

For this book, Hawking is central because he helped move cosmic origin from religious imagination into scientific investigation.

James Hartle and Stephen Hawking

Key work:
Scientific writings on the no-boundary proposal

The Hartle-Hawking no-boundary proposal suggests that the universe may be finite but without a boundary in time in the ordinary sense. The mathematics is advanced, but the philosophical importance is clear.

If the universe does not have the kind of sharp temporal edge that ordinary language imagines, then the simple idea of God starting the universe becomes less secure.

This proposal should not be exaggerated. It is not a final proof against God. But it shows how physics can challenge common-sense creator arguments.

Alan Guth

Key work:
The Inflationary Universe

Guth developed the idea of cosmic inflation, a period of extremely rapid expansion in the early universe. Inflation was proposed to address important problems in cosmology, including the flatness and horizon problems.

Inflation matters because it shows how science responds to puzzles: not by inserting God, but by building and testing deeper physical models.

Some versions of inflation also open the door to multiverse ideas, which affect the fine-tuning debate.

Andrei Linde

Key works:
Scientific writings on inflationary cosmology and eternal inflation

Linde developed influential models of inflation, including ideas related to eternal inflation. These models suggest that our observable universe may be only one region within a much larger cosmic reality.

This matters because multiverse ideas, though not proven, weaken the claim that fine-tuning leaves only one possible explanation: God.

Martin Rees

Key work:
Just Six Numbers

Rees discusses several key numbers in physics and cosmology that appear important for the existence of stars, structure, and life. His work is often discussed in relation to fine-tuning.

Rees is useful because he presents the seriousness of fine-tuning without automatically turning it into proof of God. His work helps readers understand why the question is real, but also why the answer is not simple.

Brian Greene

Key works:
The Elegant Universe
The Fabric of the Cosmos
The Hidden Reality

Greene explains relativity, quantum theory, string theory, space-time, and multiverse ideas for general readers. His books are useful because they make difficult physics more accessible.

For this book, Greene is especially helpful in showing why reality may be much stranger than common sense and why cosmology must be guided by physics rather than religious intuition.

Sean Carroll

Key works:
From Eternity to Here
The Big Picture

Carroll presents a strong naturalistic view of reality. He explains time, entropy, cosmology, physics, complexity, and meaning without invoking God.

His work is especially important for this series because he combines scientific seriousness with philosophical clarity. He shows that a natural universe can be understood without divine explanation and that meaning can exist without God.

Leonard Susskind

Key work:
The Cosmic Landscape

Susskind discusses string theory, the landscape of possible physical states, and the connection between fine-tuning and multiverse ideas.

His work matters because it provides one possible natural framework for discussing why our universe has life-permitting conditions. The landscape is debated and should not be treated as proven, but it remains important in the modern fine-tuning discussion.

Kip Thorne

Key work:
Black Holes and Time Warps

Thorne's work helps readers understand black holes, relativity, and the strange structure of space-time. These ideas are important because they show that the universe is far deeper than ordinary common sense.

For this book, Thorne is especially useful for understanding why creator arguments based on simple time and cause may be too simplistic.

Roger Penrose

Key works:
Scientific writings on singularity theorems, cosmology, entropy, and mathematical physics

Penrose's work on singularities and space-time has deeply influenced modern cosmology. He also raises important questions about the early universe, entropy, and the structure of physical law.

Penrose is useful because his work reminds us that the universe remains scientifically difficult. But difficulty is not evidence for God. It is evidence that deeper inquiry is needed.

Philosophical and Theistic Argument Sources

Thomas Aquinas

Key work:
Summa Theologica

Aquinas developed some of the most influential classical arguments for God, including arguments from motion, causation, contingency, and order.

His arguments matter because they shaped centuries of religious philosophy. But they also need careful criticism. Even if one accepts an unmoved mover or first cause, it does not automatically follow that the God of religion exists.

Aquinas represents a powerful classical form of natural theology.

Gottfried Wilhelm Leibniz

Key works:
Writings on the principle of sufficient reason and necessary being

Leibniz asked one of the deepest questions in philosophy:

Why is there something rather than nothing?

He argued that contingent reality requires an explanation and that this explanation lies in a necessary being.

His question remains powerful. But the answer "God" is not forced. A necessary foundation, if one exists, is not automatically personal, conscious, loving, or connected to any religion.

William Lane Craig

Key works:
Writings and debates on the Kalam cosmological argument

Craig is one of the best-known modern defenders of the Kalam cosmological argument. His work is important because it presents the religious case in a clear and forceful way.

He should be read to understand one of the strongest modern versions of the opposing argument. A serious science-based atheist should understand the best arguments on the other side, not only weak versions.

However, the Kalam argument still does not prove the God of religion. It moves too quickly from cause to creator, and from creator to a personal God.

Richard Swinburne

Key work:
The Existence of God

Swinburne presents a sophisticated philosophical case for theism, including arguments from order, probability, simplicity, and explanation.

His work is useful because it gives the reader a serious theistic position. But his conclusions remain open to criticism. God is not clearly simpler than the universe, and the move from cosmic order to personal God remains weak.

David Hume

Key works:
Dialogues Concerning Natural Religion
An Enquiry Concerning Human Understanding

Hume is one of the most important critics of religious arguments. His critique of design arguments remains highly relevant. He showed that moving from order in the world to a divine designer is not straightforward.

Hume matters because he understood the weakness of analogy. The universe is not a machine, house, or watch. We do not have experience of universe-makers. Therefore, arguments from human design to divine design are much weaker than they first appear.

Bertrand Russell

Key work:
Why I Am Not a Christian

Russell offered clear and influential criticisms of traditional arguments for God, including first-cause arguments. His famous response to the question of cause remains important: if everything must have a cause, then God must have a cause; if something can be uncaused, perhaps the universe can be uncaused.

Russell is useful because he expressed difficult philosophical objections in direct language.

J. L. Mackie

Key work:
The Miracle of Theism

Mackie offered a detailed philosophical critique of theistic arguments. His work is valuable for readers who want a more rigorous examination of the case for and against God.

For this book, Mackie is useful in examining cosmological, design, and fine-tuning arguments with philosophical care.

Graham Oppy

Key work:
Arguing About Gods

Oppy provides one of the most detailed philosophical critiques of theistic arguments. His work is advanced, but important for serious readers.

Oppy is especially valuable because he analyzes whether theism actually provides a better explanation than naturalism. His work supports the view that theism does not have the explanatory advantage believers often claim.

Victor Stenger

Key work:
God: The Failed Hypothesis

Stenger argues that God can be treated as a hypothesis and tested against what we observe. He strongly challenges claims from fine-tuning, cosmology, and design.

His tone can be direct, but his work is useful for this book because it focuses on whether God actually explains the universe better than science. His conclusion is that God does not.

Carl Sagan

Key works:
Cosmos
The Demon-Haunted World

Sagan combined scientific wonder with skepticism. He showed that the universe can inspire deep awe without supernatural belief.

For this series, Sagan is especially important because he demonstrates that science-based atheism need not be cold. It can be poetic, humane, morally serious, and filled with wonder.

Sagan helps show that wonder does not require worship.

Daniel Dennett

Key work:
Breaking the Spell

Dennett argues that religion itself should be studied as a natural phenomenon. While this book focuses on cosmology, Dennett's approach supports the wider series by treating belief as something that can be explained through human minds, culture, and evolution.

His work is useful for understanding how religious explanations persist even after science has made them unnecessary.

Richard Dawkins

Key work:
The God Delusion

Dawkins presents a strong public case against belief in God, including criticisms of design arguments and religious certainty.

His work is influential and useful, though the tone is often more confrontational than the tone used in this series. This series remains firm, persuasive, and evidence-based, while keeping a calmer and more humane style.

Dawkins is especially helpful on the problem of explaining the designer.

Lawrence Krauss

Key work:
A Universe from Nothing

Krauss argues that modern physics can help explain how a universe may arise from what physicists call nothing. His work is important but must be read carefully.

Critics rightly point out that physics "nothing" is not the same as philosophical absolute nothingness. This distinction matters. Still, Krauss is useful for showing that the word nothing is more complicated in physics than common sense assumes.

Official and Reference Sources

NASA Resources

NASA's public resources on cosmology, the Big Bang, dark matter, dark energy, stars, galaxies, black holes, and cosmic expansion are useful for general readers. They provide accessible summaries of scientific evidence without requiring advanced mathematics.

These resources are especially valuable for readers who want reliable, nontechnical explanations from a major scientific agency.

ESA and Planck Mission Resources

The European Space Agency's resources on the Planck mission and the cosmic microwave background are especially important for understanding precision cosmology.

Planck helped measure the early universe in remarkable detail. Its findings strengthened the evidence-based picture of cosmic history.

Stanford Encyclopedia of Philosophy

The Stanford Encyclopedia of Philosophy is useful for more advanced readers. Relevant entries include:

“Cosmological Argument”
“Fine-Tuning”
“Cosmology and Theology”
“Principle of Sufficient Reason”
“Nothingness”
“Quantum Mechanics”

These entries provide careful philosophical background and help readers understand the strongest versions of the arguments being discussed.

Internet Encyclopedia of Philosophy

The Internet Encyclopedia of Philosophy also provides useful introductory material on cosmological arguments and related philosophical topics. It is generally more accessible than some advanced academic texts.

Main Scientific Ideas Readers Should Understand

Cosmic Expansion

The universe is expanding on large scales. Distant galaxies are generally receding from us, and the farther away they are, the faster they appear to recede.

This does not mean galaxies are flying from a central explosion into empty space. It means space itself is expanding.

Cosmic expansion is one of the foundation stones of modern cosmology.

Redshift

Light from distant galaxies is stretched toward longer wavelengths as the universe expands. This is called cosmological redshift.

Redshift helped scientists understand that the universe is expanding and that the cosmos has a physical history.

Cosmic Microwave Background Radiation

The cosmic microwave background is the cooled radiation left from the early universe, when the universe became transparent and light could travel freely.

It is one of the strongest pieces of evidence for Big Bang cosmology.

It is not the voice of God.

It is the afterglow of a hot early universe.

Big Bang Cosmology

Big Bang cosmology describes the observable universe expanding and cooling from an earlier hot, dense state.

It is not an explosion into empty space.

It is not the same as religious creation.

It is not proof of God.

It is a scientific model supported by evidence.

Dark Matter

Dark matter is inferred from its gravitational effects, especially in galaxies and large-scale cosmic structure. Its exact nature remains unknown.

Dark matter is a real scientific mystery.

But mystery is not evidence for God.

Dark Energy

Dark energy is the name given to whatever is driving the accelerated expansion of the universe. Its deeper nature remains uncertain.

Like dark matter, it shows that science has more to learn.

It does not show that God exists.

Inflation

Inflation is the idea that the early universe underwent an extremely rapid expansion. It was proposed to explain certain features of the universe, such as its large-scale uniformity and near-flat geometry.

Some versions of inflation suggest multiverse possibilities.

Inflation is important, but not all questions about it are settled.

Fine-Tuning

Fine-tuning refers to the observation that certain features of the universe appear to fall within ranges that allow stars, chemistry, complexity, planets, and life.

It is a serious issue.

But it is not proof of God.

Possible responses include deeper physical law, selection effects, multiverse ideas, unknown physics, or brute fact.

Anthropic Principle

The anthropic principle says that our observations are conditioned by our existence as observers. We can observe only a universe compatible with observers.

This weakens simple design arguments by reminding us that we should not be surprised to find ourselves where beings like us can exist.

Multiverse

The multiverse is the idea that our observable universe may not be all that exists. There may be many cosmic domains, perhaps with different conditions.

The multiverse is not proven.

But it shows that divine design is not the only possible response to fine-tuning.

Quantum Cosmology

Quantum cosmology applies quantum ideas to the universe as a whole, especially the earliest universe.

It remains incomplete and difficult.

But it shows that ordinary ideas of cause, before, beginning, and nothing may not apply simply to cosmic origin.

No-Boundary Proposal

The Hartle-Hawking no-boundary proposal suggests that the universe may be finite but without a sharp boundary in time in the ordinary sense.

It does not disprove God by itself, but it challenges the simple idea that the universe needs a divine starter at a first moment.

Main Philosophical Ideas Readers Should Understand

First Cause

The first-cause argument says the universe must have a cause, and that cause is God.

The problem is that even if a first cause exists, it does not automatically become a personal, loving, prayer-answering God.

Kalam Cosmological Argument

The Kalam argument says:

Whatever begins to exist has a cause.

The universe began to exist.

Therefore, the universe has a cause.

This argument is influential, but it depends on applying ordinary causation to the universe as a whole. It also does not establish the God of religion.

Principle of Sufficient Reason

This principle says that everything must have a sufficient explanation for why it exists or why it is the way it is.

The principle raises deep questions, but it does not automatically prove God. It may be that some ultimate features of reality are brute facts, or that the deepest explanation is natural rather than divine.

Necessary Being

Some philosophers argue that God exists necessarily.

But calling God necessary does not prove God exists. And even if necessary reality exists, it is not automatically personal or religious.

Brute Fact

A brute fact is something that has no deeper explanation, or at least no deeper explanation within a given framework.

The believer may stop with God. The atheist may stop with reality itself, or with a deeper natural foundation. The question is which stopping point is more justified by evidence.

God of the Gaps

The God of the gaps means inserting God where knowledge is incomplete.

This has repeatedly failed. Many things once explained by God later received natural explanations.

The origin of the universe should not be treated as a final hiding place for God.

Suggested Reading Path for General Readers

Readers who want a broad and accessible introduction to cosmology and science-based wonder may begin with:

Carl Sagan, *Cosmos*

Stephen Hawking, *A Brief History of Time*

Steven Weinberg, *The First Three Minutes*

Sean Carroll, *The Big Picture*

Brian Greene, *The Fabric of the Cosmos*

Martin Rees, *Just Six Numbers*

Readers who want to understand fine-tuning, multiverse ideas, and cosmic design arguments may read:

Martin Rees, *Just Six Numbers*

Leonard Susskind, *The Cosmic Landscape*

Brian Greene, *The Hidden Reality*

Victor Stenger, *God: The Failed Hypothesis*

Stanford Encyclopedia of Philosophy, "Fine-Tuning"

Readers who want to understand philosophical arguments for and against God may read:

David Hume, *Dialogues Concerning Natural Religion*

Bertrand Russell, *Why I Am Not a Christian*

J. L. Mackie, *The Miracle of Theism*

Graham Oppy, *Arguing About Gods*

Richard Swinburne, *The Existence of God*

William Lane Craig, writings on the Kalam cosmological argument

Readers who want to explore quantum cosmology and the origin question more deeply may read:

Stephen Hawking, *A Brief History of Time*

Stephen Hawking and Leonard Mlodinow, *The Grand Design*

Sean Carroll, *From Eternity to Here*

Brian Greene, *The Fabric of the Cosmos*

Kip Thorne, *Black Holes and Time Warps*

How to Read Theistic Sources

A science-based atheist should not avoid the strongest religious arguments.

It is useful to read writers such as William Lane Craig and Richard Swinburne because they present serious theistic cases. Reading them helps us avoid weak criticism and straw-man arguments.

However, readers should ask careful questions:

Does the argument prove a personal God?

Does it prove any specific religion?

Does it give evidence, or only possibility?

Does it explain more than it assumes?

Does it answer who created God?

Does it provide a mechanism?

Does it make testable predictions?

Does it improve scientific understanding?

Often, the answer is no.

Theistic arguments may sound impressive, but they frequently move from mystery to God too quickly.

How to Read Scientific Sources

Scientific sources should also be read carefully.

Science is powerful, but not finished.

Readers should avoid turning science into dogma. The Big Bang model is strong, but not the final answer to every cosmic question. Inflation is important, but still debated. The multiverse is possible, but not proven. Quantum cosmology is fascinating, but incomplete. Dark matter and dark energy remain mysterious.

This honesty does not weaken science-based atheism.

It strengthens it.

The case for science-based atheism does not depend on pretending that science knows everything. It depends on recognizing that science is the best method we have for discovering how reality works, and that God has not been shown to explain what science does not yet know.

A Note on Strongly Critical Books

Some modern books argue strongly against religion and belief in God. They can be useful, especially when examining design arguments, the God of the gaps, and the problem of explaining the designer.

However, tone matters.

The purpose of this series is not to insult believers. It is to guide readers toward science-based, evidence-based atheism through reason, evidence, history, and calm persuasion. A book can be firm without being crude. It can be direct without being careless. It can reject God-based explanations without dehumanizing religious people.

For that reason, strongly critical authors should be read for their arguments, but not necessarily imitated in style. The best approach for this series is calm confidence, intellectual fairness, and steady persuasion.

Final Note

The universe is not simple.

It is vast, ancient, dynamic, lawful, strange, and deeply mysterious.

But mystery is not evidence for God.

The thinkers and sources listed in this appendix show that cosmology has become a serious evidence-based science. The universe can be studied through light, radiation, gravity, mathematics, observation, and physical law. Its history can be reconstructed. Its structure can be measured. Its mysteries can be investigated.

Religion once claimed the sky.

Then the sky became astronomy.

Religion claimed creation.

Then cosmic history became physics.

Religion claimed the beginning.

Then science began to examine even that.

The story is not finished. Science still has much to learn. But the direction is clear.

God is not needed to explain the universe.

A creator is not required.

The cosmos is not less magnificent because it is natural.

It is more magnificent because it is real.

No God is required where reality can explain itself.

Thank You and Please Review the Book

Thank you for taking the time to read this book. I sincerely hope it has encouraged you to think more deeply, question more honestly, and look at religion, science, and human belief with greater clarity.

My goal in writing the **No God Required Series** is simple: to present serious, science-based, and easy-to-understand ideas about God, religion, nature, morality, and human meaning. These books are written for readers who value evidence, reason, intellectual honesty, and the courage to ask difficult questions.

If you found this book thoughtful, helpful, or worth reading, I would be truly grateful if you could leave a brief and honest review. Even a few sincere lines can make a meaningful difference. Your review helps other readers discover this series and encourages more open, rational, and evidence-based discussion about some of life's deepest questions.

Thank you again for your time, your trust, and your support.

Warm regards,
Dr. Prabhat Das

OTHER BOOKS IN THIS SERIES

No God Required: How Science Explains Reality Without the Divine is a bold, clear, science-based series by **Dr. Prabhat Das**. It reveals why humans created gods, how science made supernatural explanations unnecessary, and how meaning, morality, courage, and wonder can flourish without the divine.

The 7 Books in This Series: -

Book 1: **Why and How Humans Created God**

Book 2: **How Science Made God Unnecessary**

Book 3: **The Universe Without a Creator**

Book 4: **Life Without a Designer**

Book 5: **Mind Without Soul, Morality Without God**

Book 6: **The Failure of the Personal God**

Book 7: **The Scientific Case for Atheism**